Loa loa: Latest Advances in Loiasis Research

Jean Paul Akue
Editor

Loa loa: Latest Advances in Loiasis Research

Springer

Editor
Jean Paul Akue
Parasitology Dept.
CIRMF
Franceville, Gabon

ISBN 978-3-031-49449-9 ISBN 978-3-031-49450-5 (eBook)
https://doi.org/10.1007/978-3-031-49450-5

This Springer imprint is published by the registered company Springer Nature Switzerland AG
The registered company address is: Gewerbestrasse 11, 6330 Cham, Switzerland

To my wonderful love:
Priscille Assa in memory, Claire d'Assise Oyane Nzamba,
To my children:
Khevane, Jeanis Paule, Amandine, Junior Akue, Bienvenue, Melchy Akue
To my Grandchildren:
Priscille Aloume, J. J Akue -Assengone, Matty O' Claine

Preface

Loa loa is endemic in west and central Africa, where more than 200 hundred million people are at risk (i.e., exposed to infective bites of Chrysops vectors of *Loa loa*) and about 13 million are infected. The parasite has been neglected to date due to the relative rarity of significant clinical manifestations in infected individuals living in endemic areas. However, the explosion of side effects due to mass control programs for other parasites has drawn attention to the parasite *Loa loa*. Now it is acknowledged that *Loa loa* has a significant impact on public health with disability-adjusted life years (DALYs) of 412.9 DALYs per 100,000 individuals (Veletzky et al. 2020). Increased mortality has also been reported in hypermicrofilaremic individuals (Hemilembolo et al. 2023). Furthermore, a study has shown that some clinical manifestations credited to other pathogens, such as fever, headache, myalgia, paresthesia, and renal disease, are more likely due to *Loa loa* infection (Veletzky et al. 2020).

Although definitive diagnosis relies on the detection of microfilariae in blood, subconjunctival passage of the adult worm (eyeworm) is considered pathognomonic. Despite the high specificity of blood microfilaremia and eyeworm, these are insensitive for the diagnosis of *Loa loa*. New tests based on biomarkers have been developed using loop-mediated isothermal amplification (LAMP), *polymerase chain reaction* (PCR), enzyme-linked immunosorbent assay (ELISA) (Gobbi et al. 2020), and lateral flow assays (LFA) (Pedram et al. 2017). A new less tedious technique for counting blood microfilariae (LoaScope) has also been developed (D'ambrosio et al. 2015). Simultaneous major advances in treatment have been lacking. Ivermectin, DEC, and albendazole are still in use with special protocols for complicated cases. The potential for fatal side effects with DEC or ivermectin treatment has prompted interest in ways to overcome this pathology. In this regard, it is interesting to look at immunity against *Loa loa*. One observation that suggests a role for immunity in the clinical response to infection is the spectrum of presentations of *Loa loa* infection, which can be divided into four (4) groups: (a) those with microfilariae in the blood (microfilaremic), (b) those with detectable filarial DNA in their blood but without detectable blood microfilariae (cryptic infection), (c) those with ocular passage of an adult worm or symptomatic without a history of eye worm, but no detectable blood microfilariae (amicrofilaremic), and (d) individuals exposed to infective bites but without any sign of infection (endemic control group). Whether human host or parasite genetics influence this clinical spectrum is unknown. However, several studies suggest active involvement of the immune response; for example, the appearance of microfilariae in the peripheral blood coincides with downregulation of Th2 cells through the induction of anergy with a reduction in Th2 cell proliferation but increased IgG4 levels. The elevation of IgG4 occurs concurrently with elevated IgE, suggesting cross-reactivity between the two Ig isotypes with the potential end result of reduced allergic reaction. Additionally, the existence of amicrofilaremic and endemic control groups suggests some protective immunity acting either on microfilariae or on incoming infective stage larvae (L3).

The probable effector mechanism leading to reduced circulating parasites has been described as antibody-mediated cell cytotoxicity (ADCC). While some target antigens from *Loa loa* have been identified, the nature and structure of such antigens have not been fully elucidated to date. Although experimental models have been developed, the lack of abundant sources for crude antigen preparations to study potential diagnostic or immunoprophylactic tools has led to the use of recombinant DNA technology to generate material for such studies. Molecular knowledge of *Loa loa* may uncover principal pathways in the parasite metabolism that could lead to the development of diagnostic tools, new drugs, or vaccines.

The objective of this book is to highlight the real face of the neglected *Loa loa* parasite. This book will provide insights for policy makers, health workers, students, and researchers all over the world where the disease was unknown but became emergent due to the fast mobility of the world population.

We believe that this book will help improve the point-of-care treatment of *Loa loa* and will be a learning tool for students and a way to go for policy makers and researchers.

References

D'ambrosio M, Bakalar M, Bennuru S, Reber C, Skandarajah A, Nilsson L, Switz N, Kamgno J, Pion S, Boussinesq M, Nutman TB, Fletcher DA (2015) Point-of-care quantification of blood-borne filarial parasites with a mobile phone microscope. Sci Transl Med 7(286):286re4. https://doi.org/10.1126/scitranslmed.aaa3480

Hemilembolo Marlhand C, Niama AC, Campillo JT, Pion SD, Missamou F, Whittaker C, Kankou J-M, Ndziessi G, Bileckot RR, Boussinesq M, Chesnais CB (2023) Excess mortality associated with loiasis: confirmation by a new retrospective cohort study conducted in the Republic of Congo. Open Forum Infect Dis 10(3):ofad103. https://doi.org/10.1093/ofid/ofad103. eCollection 2023 Mar

Gobbi F, Buonfrate D, Boussinesq M, Chesnais CB, Pion SD, Silva R et al (2020) Performance of two serodiagnostic tests for loiasis in a Non-Endemic area. PLoS Negl Trop Dis 14(5):e0008187. https://doi.org/10.1371/journal.pntd.0008187

Pedram B, Pasquetto V, Drame PM, Ji Y, Gonzalez-Moa MJ, Baldwin RK et al (2017) A novel rapid test for detecting antibody responses to *Loa loa* infections.

PLoS Negl Trop Dis 11(7):e0005741. https://doi.org/10.1371/journal.pntd.0005741

Veletzky L, Hergeth J, Stelzl DR, Mischlinger J, Manego RZ, Mombo-Ngoma G, McCall MB, Adegnika AA, Agnandji ST, Metzger WG, Matsiegui PB, Lagler H, Mordmüller B, Budke C, Ramharter M (2020) Burden of disease in Gabon caused by loiasis: a cross-sectional survey. Lancet Infect Dis 20:1339–1346. https://doi.org/10.1016/S14733099(20)30256

Franceville, Gabon Jean Paul Akue

Acknowledgments

We sincerely thank all the authors with humility. Without your unconditional support, this book never would have come to be. Thank you for your willingness to share tremendous acknowledgment on *loa loa*. This goes to Amy Klion, Wolfram Meztger, Luzia Veletzky, Michel Boussinesq, Samuel Wanji, Kamgno Joseph, and their team. We particularly thank Amy and Luzia who took the time out of their very busy schedules to read an advance copy of this book and write constructive comments. One part of the work on *Loa loa* was supported by CIRMF, sponsored by Total Gabon and Gabonese state.

Contents

Contributors

Jean Paul Akue Neglected Parasitosis, CIRMF, Franceville, Gabon

Michel Boussinesq Institut de Recherche pour le Développement, Unité Mixte Internationale (UMI) 233, Institut National de la Santé et de la, Recherche Médicale (INSERM) U1175, Université de Montpellier, Montpellier Cedex, France

Roland Dieki Neglected Parasitosis, CIRMF, Franceville, Gabon

Arnauld Efon Ekangouo Higher Institute for Scientific and Medical Research (ISM), Yaoundé, Cameroon

Fanny Fri Fombad Parasite and Vector Biology Research Unit, Department of Microbiology and Parasitology, Faculty of Science, University of Buea, Buea, Cameroon
Research Foundation in Tropical Diseases and the Environment, Buea, Cameroon

Joseph Kamgno Higher Institute for Scientific and Medical Research (ISM), Yaoundé, Cameroon
Department of Public Health, Faculty of Medicine and Biomedical Sciences, University of Yaoundé I, Yaoundé, Cameroon

Amy Klion Human Eosinophil Section, Laboratory of Parasitic Diseases, National Institutes of Allergy and Infectious Diseases, National Institutes of Health, Bethesda, MD, USA

Line Edwige Mengome Institut de Pharmacopée et de Médecine Traditionnelle (IPHAMETRA - CENAREST), Libreville, Gabon

Wolfram G. Metzger Faculty of Medicine, Eberhard Karls University Tübingen, Tübingen, Germany

Hugues C. Nana Djeunga Higher Institute for Scientific and Medical Research (ISM), Yaoundé, Cameroon

Narcisse Gandjui Parasite and Vector Biology Research Unit, Department of Microbiology and Parasitology, Faculty of Science, University of Buea, Buea, Cameroon
Research Foundation in Tropical Diseases and the Environment, Buea, Cameroon

Samuel Wanji Parasite and Vector Biology Research Unit, Department of Microbiology and Parasitology, Faculty of Science, University of Buea, Buea, Cameroon
Research Foundation in Tropical Diseases and the Environment, Buea, Cameroon

Valerine Chunda Parasite and Vector Biology Research Unit, Department of Microbiology and Parasitology, Faculty of Science, University of Buea, Buea, Cameroon
Research Foundation in Tropical Diseases and the Environment, Buea, Cameroon

Luzia Veletzky Department of Medicine I, Division of Infectious Diseases and Tropical Medicine, Medical University of Vienna, Vienna, Austria

Linda Djune Yemeli Higher Institute for Scientific and Medical Research (ISM), Yaoundé, Cameroon

The History of the *Loa loa* Parasite, Its Biology and Experimental Models

Valerine Chunda, Fanny Fri Fombad, Narcisse Gandjui, and Samuel Wanji

Abstract

Loiasis affects an estimated population of 14.4 million people living in rainforest areas of West and Central Africa. Although the disease is known to be mild, it has significantly hampered the goals' achievement of the Onchocerciasis elimination program as individuals with high microfilarial loads may suffer from severe adverse events post-ivermectin treatment for Onchocerciasis in co-endemic regions. In the past, preclinical research on loiasis was extremely challenging as it was limited to nonhuman primates. However, recent advances in *Loa loa* research have addressed this issue by establishing novel experimental tools which have facilitated the advancement of loiasis research. This chapter summarizes available information on the history of *Loa loa*, its biology and current experimental models (most of which were developed by our research group) that have advanced research on loiasis, bringing in new opportunities for understanding the pathogenesis of the diseases and investigating for new treatment options.

Keywords

Loa loa · History · Biology · In vitro systems · Animal models · Drug screening

V. Chunda · F. F. Fombad · N. Gandjui · S. Wanji (✉)
Parasite and Vector Biology Research Unit, Department of Microbiology and Parasitology, Faculty of Science, University of Buea, Buea, Cameroon

Research Foundation in Tropical Diseases and the Environment, Buea, Cameroon

J. P. Akue (ed.), *Loa loa: Latest Advances in Loiasis Research*,
https://doi.org/10.1007/978-3-031-49450-5_1

1 The History of *Loa loa* Parasite

The first definitive case of *Loa loa* was recorded by a French surgeon Mongin (1770) who failed in an attempt to remove the adult worm passing across the eye of a woman in Santa Domingo, in the Caribbean. Years later Bajon (1777) reported a similar case observed in a young girl from Cayenne. The first description of the parasite was documented in 1778 by Guyot in Angola, and he termed it "Loa" meaning "worm." Thereafter, there have been increasing numbers of similar records, such as Lorenz (1890), Manson (1904), and Stiles (1905), who coined the definitive term "*Loa loa*." Manson discovered the microfilariae of *L. loa* in the blood (1891) and named it *Filaria diurna* referring to the periodicity of the parasite. Broden and Rodhain (1908) saw microfilariae in cerebrospinal fluid, while Kulz (1908) described the clinical symptoms of the disease. Van Campenhoot (1900) established the relationship between edema and *L. loa* in several patients. These observations were also made by Mouchet in Leopoldville, while in 1912, the British parasitologist Robert Thomson Leiper confirmed the development of *L. loa* in the tabanid flies *Chrysops silacea* and *Chrysops dimidiata* (Dieki et al. 2022). Kleine (1915) and the Connals (1922, 1923) drew attention to the sites occupied by the developing larvae within the *Chrysops* flies.

2 The Biology of Filaria *Loa loa* in the Mammal Host and the Vector

2.1 Taxonomic Classification of *Loa loa*

L. loa belongs to the following (Schmidt and Roberts 2006):

Kingdom: Animalia
Phylum: Nematoda
Class: Phasmidea
Order: Spirurida
Superfamily: Filarioidea
Family: Onchocercidae
Genus: *Loa*
Species: *Loa loa*

2.2 The Pathogen *Loa loa*

Adult *L. loa* live and migrate into the subcutaneous and deep connective tissues of their hosts usually lying between fascia and muscle layers. There they produce large numbers of embryonic progeny called microfilariae (Whittaker et al. 2018). They are sometimes seen passing under the conjunctiva of the eye hence the name "eye worm." Unlike other filariae, *L. loa* lacks the obligate intracellular *Wolbachia*

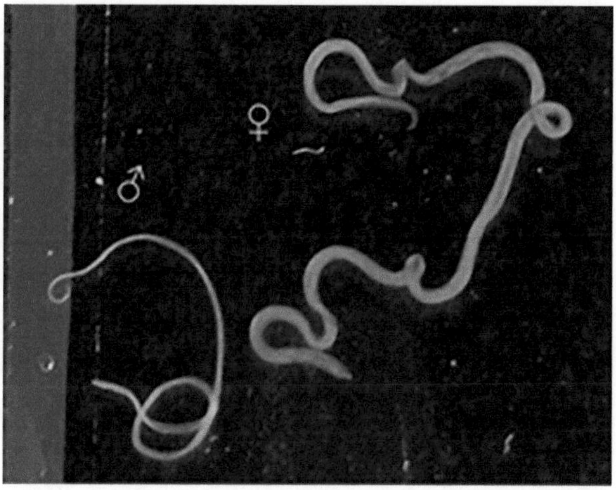

Fig. 1 Photograph of female and male *Loa loa* (Pionnier et al. 2019)

endosymbionts and does not have any specialized metabolic route as an alternative (Desjardins et al. 2013). Both male and female adults *Loa* worms are made up of a simple head with no lips and eight cephalic papillae, a long, slender body and a blunt tail. The cuticle is covered with irregular, small bosses, except at the head and tail. In most cases, the females are longer and wider in diameter than their male counterparts (Fig. 1). The females measure 40–70 mm long and 0.5 mm wide with a vulva of about 2.5 mm from the anterior end. Males are about 30–34 mm long and 0.35–0.43 mm wide (CDC 2008). The microfilariae circulate in the peripheral blood during the day and reach their maximal concentration between 10:00 and 15:00 hours. They are sheathed and are about 250–300 μm long and 5–8 μm wide. They have body nuclei that are continuous from the head to the tail.

2.3 The Vectors of *Loa loa*

Microfilariae of *L. loa* are transmitted by several species of tabanid flies (order: Diptera). Although horseflies of the *Tabanus* genus are often mentioned as *Loa* vectors, the two prominent vectors are from the *Chrysops* genus of tabanids (*Chrysops silacea* and *C. dimidiata*), Fig. 2. *C. dimidiata* and *C. silacea* live in forested and muddy habitats like swamps, streams, and reservoirs and in rotting vegetation and are particularly attracted by smoke and blue tissue. These species exist only in Africa and are popularly known as deerflies, mango flies, or mangrove flies. *Chrysops* species are 5–20 mm long with a large head and downward-pointing mouthparts. Their wings are clear or speckled brown. Female flies require a fair amount of blood for their reproductive purposes and thus may take multiple blood meals from the same host if disturbed during the first one. The batch of eggs is deposited near water, mud, or leaves where the eggs hatch in 5–7 days (Padgett and Jacobsen 2008).

Fig. 2 Picture of *Chrysops* species, vector of *L. loa* (**a**: *C. silacea*, **b**: *C. dimidiata*) (source: Schmidt and Roberts 2006)

The larvae (1–6 cm long) mature in water or soil, where they feed on organic material such as decaying animal and vegetable products taking a year before they pupate, with probably seven molts. The pupa is partially buried, and the adult emerges after 1–3 weeks. When fully mature, *C. silacea* and *C. dimidiata* assume the day-biting tendencies of all tabanids (Padgett and Jacobsen 2008). The bite of the mango fly can be very painful, possibly due to the laceration style employed. Rather than puncturing the skin like a mosquito does, the fly makes a laceration in the skin and subsequently laps up blood.

2.4 Simian Loiasis

Several monkey species have been found naturally infected with *L. loa* sp. in various African countries (Orihel and Moore 1975). *L. loa* has been found in: *Mandrillus leucophaeus*, *Cercopithecus nictitans martini,* and *Ce. mona mona* in Kumba, South West region of Cameroon (Duke and Wijers 1958), *Mandrillus sphinx* (Akue et al. 2001) and in *Papio* species (baboons), (Orihel and Moore 1975; Orihel and Eberhard 1985). Of all the simian hosts, *Mandrillus leucophaeus* has been found to have the highest prevalence (96%) of simian loiasis; the highest numbers of adult worms with the adults and mf being longer than those from other monkeys. This is probably because the parasite is well adapted to this monkey species (Duke and Wijers 1958). The mf of the simian form *L. loa* exhibits nocturnal periodicity with a peak corresponding to the hours of activity of the vectors. In Kumba, Cameroon, the vectors are *Chrysops langi* and *Chrysops centurionis*, which are canopy-dwelling and bite only between 17:00 and 21:00 hours (Duke 1955). Little is known about the pathogenicity of *L. loa* in naturally infected monkeys, and no clinical signs of infection have ever been reported in these hosts (Pinder et al. 1994). Studies conducted by Duke (1964) showed that hybridization of the human and simian forms of *Loa loa* might be possible and that the hybrids are fertile with the two forms belonging to the same species. Given the predominantly diurnal periodicity of the hybrid mf, the fact that the adult worms were less fertile than their parents, and the vectors of the forms showed very different biting habits, it seems likely that hybridization rarely occurs under natural conditions and that simian and human parasites are in the process of divergent evolution (Duke 1964). It is not known if simian *L. loa* species can infect humans. The only reported attempt to induce such an infection experimentally was unsuccessful (Duke 2004).

2.5 Parasite Life Cycle

The life cycle of *L. loa* is relatively simple, and the parasite has different life cycle stages which include the microfilaria, the larval stages (2, 3, 4), and the adult stage. The life cycle involves both the fly vector (intermediate host) and the human (definitive host). The L4, adult worm (L5), and L1 (mf) are found in the human host, whereas the L2 and L3 are found in the vector. The mf ingested by a *Chrysops* during a blood meal on an infected individual develops in the fly. For the first 24 h, the parasites are located in cells of the fat body in the thorax and, especially the abdomen of the fly. Growth of parasites causes the cells to burst. When mature, some 10–12 days (Fig. 3) after the blood meal, the L3 (infective stage) migrates from the

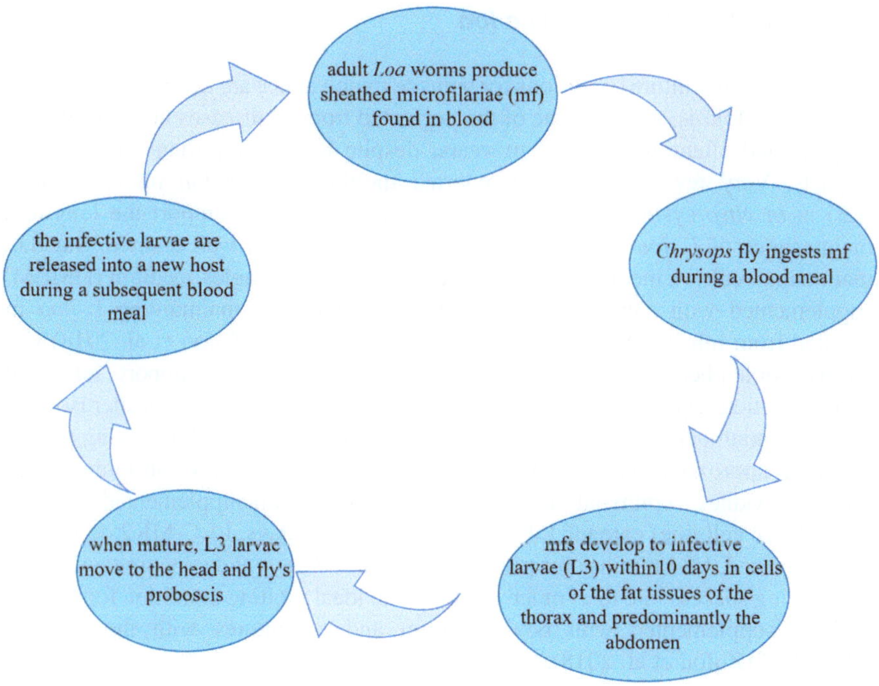

Fig. 3 Life cycle of *L. loa*

thorax/abdomen to the head of the insect (Whittaker et al. 2018). During a subsequent blood meal, the L3 escapes from the proboscis of the insect and is deposited on the surface of the potential host's skin. Some of them succeed to enter the skin, via the tunnel cut by the vector, and reach the dermis where they rapidly begin a migration via the host's lymphatic system (Bain et al., 1998). The third molt, from L3 to L4, begins as from day 9 post-infection, whereas the fourth molt to the young adult occurs by day 19 (Bain et al. 1998). The adult worms live between the layers of loose connective tissue under the skin and between the fascial layers underlying the somatic muscles where they reached maturity (Sadia et al. 2008). Their mean life span is unknown but can live up to 20 years. Once fertilized, the female worms begin releasing thousands of mf each surrounded by a sheath that is a relict of the eggshell. The released mf passes into the host's lymphatic system before accumulating in peripheral blood periodically. The pre-patent interval (the interval between the penetrations of L3 into the dermis of the new host and the first occurrence of mf in the peripheral blood) is between 4 and 5 months (Orihel and Moore 1975; Wanji et al. 2015; Pionner et al. 2019). When the *Chrysops* vector takes a blood meal from a microfilaremic host, the vector becomes infected and the life cycle begins again.

3 Experimental Models of *Loa loa* Filaria

3.1 In Vitro Model of *Loa loa*

Suitable in vitro culture conditions for parasite maintenance are needed to foster drug research for loiasis, which is one of the neglected tropical diseases that has attracted only limited attention over recent years, despite having important public health impacts. Very few authors have reported molting of *Loa loa* infective larvae (L3) in *in vitro* systems. Optimum culture conditions that support the long-term maintenance of *L. loa* mf or developing larvae have been well defined. Standard mammalian culture media, such as Dulbecco's modified Eagles medium (DMEM), supplemented with calf serum (5–15%) are sufficient in maintaining *L. loa* mf purified from infected baboon blood for about 6–12 days (Zofou et al. 2018) with a fully motile phenotype. Similarly, calf serum supplementation supports full motility of infectious stage L3 larvae in mammalian culture for up to 8 days after isolation from the mouthparts of *Chrysops* flies. Completion of the L3 to L4 molting process occurs in approximately 20% of these basic cultures. Addition of an immortalized monkey kidney epithelial cell monolayer to serum-supplemented cultures (LLC-MK2 cell line) extends full motility of mf to 22 days. LLC-MK2 co-cultures also increase L3/L4 motility for periods as long as 17 days and concomitantly boost the molting success rate to as much as 60%. The ideal molting condition for *L. loa* is DMEM supplemented with bovine serum and co-cultures with the cell line LLC-MK2 (Zofou et al. 2018).

3.2 In Vivo Models of *Loa loa*

3.2.1 Nonhuman Primate Models

The *L. loa* nonhuman primate (NHP) model is one of the best-studied animal models of human filarial infections. Much of the biology of *L. loa* has been studied using NPH (Wahl and Georges 1995). The human strain of *L. loa* can be experimentally transmitted to *Mandrillus leucophaeus* (drill), *M. sphinx* (mandrill), *Papio anubis* (baboon), *Erythrocebus patas* (patas monkey), and *Macaca mulatta* (rhesus macaque) (Duke and Wijers 1958; Orihel and Moore 1975; Orihel and Eberhard 1985; Pinder et al. 1994). Chimpanzee (*Pan troglodytes*) seems to be non-susceptible to the parasite. The pre-patent interval after experimental infection of susceptible monkeys is about 150 days, irrespective of the species of experimental host involved. In most monkey species, once the infection has become patent, the microfilarial densities increase sharply, reach a peak, and then fall within several weeks to very low levels, which then persist throughout the infection (Duke 1960a). The spleen plays a major role in the clearance of mf from the peripheral blood (Duke 1960a; Wanji et al. 2015, 2017). Splenectomized monkeys develop very high *L. loa* microfilaremia which can persist for many months (Duke 1960b; Orihel and Eberhard 1985; Wanji et al. 2015). Although drills and monkeys are excellent laboratory hosts for *L. loa*, ethically, drills are under strict restrictions according to the Convention on International Trade in Endangered Species (CITES) classification of primates. As such, this species is no longer used for biomedical research. As an alternative, the baboon offers potential to be used as an experimental NHP model for *L. loa* as the parasite behaves in this primate in essentially the same way as it does in the drill (Orihel and Moore 1975). The pre-patent interval after experimental infection of splenectomized baboons is 5 months, with microfilaremia accruing for periods up to 18 months (Wanji et al. 2015). Importantly for the study of pathophysiology of microfilaricidal adverse reactions, hyper-microfilaremia (>30,000 mf/mL) can be achieved in the majority (approximately 70%) of infected splenectomized baboons and all infected animals develop eosinophilia significantly exceeding the normal range (Wanji et al. 2015).

3.2.2 Mice Models

While closely emulating the life cycle of human loiasis, throughput of the baboon NHP model is severely constraining for anti-filarial drug research and to identify potential targets for adjunct therapies to limit post-ivermectin adverse reactions. Availability of *L. loa* susceptible laboratory rodent models, particularly mice, would be a step-change improvement, both because they are a convenient, standardized model with tractable genetic and immunological tools available but also as a less sentient animal substitute to reduce or replace usage of NHP (Wanji et al. 2021). *L. loa* does not undergo a full course of development in laboratory "wild-type" immunocompetent mice. *L. loa* infective larvae administered subcutaneously survive only for a week in BALB/c mice, an inbred laboratory strain that is conversely permissive to other filariae such as *Litomosoides sigmodontis* and *Brugia* spp., respectively (Petit et al. 1992; Turner et al. 2018). When BALB/c mice are immuno-suppressed with hydrocortisone, *L. loa* survival is extended for up to 3 weeks (Bain et al. 1998; Wanji et al. 2002). Control of infection in BALB/c mice is associated with a "type 2" cellular immune response of splenocytes when restimulated with *L. loa* L3 antigen, notably with elevated production of interleukin-4, interleukin-9 and, interleukin-13 (Chunda et al. 2020). Confirming a role for both IL-4/IL-13 signaling and IL-5 in the early adaptive immune control of *L. loa* in mice, BALB/c IL-4 receptor and IL-5 combination-deficient mice are susceptible to pre-patent adult *L. loa* infections (Tendongfor et al. 2012). The use of C-C chemokine receptor (CCR)-3 mice, deficient in recruitment of eosinophils via eotaxins and other eosinophil chemokines, demonstrates extended survival of *L. loa* developing larvae. Linking type-2 immune responses with tissue eosinophil recruitment as a mediator of early immunity to loiasis (Fombad et al. 2019). Since the minimum pre-patent period prior to the release of mf in blood is 5 months, Pionnier et al. investigated the long-term parasitological success of *L. loa* infection in a panel of "severe combined" lymphopenic immunodeficient mice lacking all adaptive immunity and facets of innate immune responses (Pionnier et al. 2019). Moderate levels of pre-patent adult *L. loa* infection were evident in CB.17 SCID mice (a BALB/c congenic background strain) at 3 months post-infection; meanwhile, non-obese diabetic (NOD) SCID mice and BALB/c RAG2−/− mice had cleared the infection at the same time point. Fecund adult *Loa* infections in the natural parasitic niche were reproducibly evident at 5 months in compound immunodeficient, lymphopenic mouse strains: NOD SCID γc−/− and BALB/c RAG2−/−γc−/−, which lack both lymphocytes and the common gamma chain (γc) cytokine signaling pathway. At this time point, in both compound immunodeficient mouse strains, most worms were found in the natural tissue niches of *Loa* adult stages, namely the subcutaneous and muscle fascia tissues, while some adults were recovered from cardiopulmonary tissues as well as the pleural and peritoneal cavities. Parasitism of *L. loa* within these organs suggests that larvae might have migrated via the lymphatics through the thoracic duct, corroborating a theory that filariae have a unified lymphatic larval phase (Bain et al. 1994; Kilarsk et al. 2019). By implanting defined a number of male and female *L. loa* adults from these strains under the skin of either BALB/c RAG2−/− or RAG2−/−γc−/− recipients, it was possible to establish microfilaremic mice 1-month post-implant with high rates of adult worm survival retained in subcutaneous tissues.

Due to the long pre-patent period following L3 infection, infusing purified *L. loa* mf derived from infected baboon NHP directly into venous blood of mice is easier (Pionnier et al. 2019). This approach was based on the success of establishing long-term microfilaremia in a variety of inbred mouse strains and genetic knockouts using the human lymphatic filarial parasite, *Brugia malayi* (Cadman et al. 2014; Pionnier et al. 2020). After infusion with inoculates of 40,000 purified mf, persistent microfilaremia could be established in either BALB/c or CB.17 SCID mice over a period of at least 8 days. A good number of mf were sequestered in cardiopulmonary circulation (approximately 70% of inoculates) with a scant peripheral microfilaremic evident. Both peripheral microfilaremia and cardiopulmonary microfilaremia were slightly improved in splenectomized BALB/c mice although non-splenectomized

SCID mice supported the highest parasitaemia, indicative of adaptive immunity regulating density of *L. loa* microfilaremia. Increasing the *L. loa* inoculate to 100,000 mf per mouse leads to significantly increased cardiopulmonary microfilaremia in both SCID and BALB/c mice. The bias of *L. loa* mf accumulation in the cardiopulmonary circulation may reflect the anatomical differences between murine and human microvasculature. In addition, because in humans *L. loa* exhibits a diurnal periodicity (Hawking 1955), physiological cues for peripheral circulatory migration versus cardiopulmonary sequestration may vary between mice and humans. Human sub-periodic *B. malayi* also demonstrates a tropism for cardiopulmonary circulation when infused into mice (Halliday et al. 2014; Sjoberg et al. 2019). It has been demonstrated in follow-up experiments that mf purified from blood samples of infected human volunteers was found to be comparable to those isolated from baboons and was used as a more abundant and ethical source to avoid NHP usage in onward applications of the mouse models.

3.3 Experimental Generation of *Loa loa* Infective Larvae

The successful development of a range of loiasis *in vivo* models increases opportunity for translational science applications. However, the generation of infectious stage larvae has been limited to dissections of wild-caught *Chrysops* in baited traps or human landing catch, which is laborious and hinders throughput. Recently, it was demonstrated that *L. loa* mf purified from experimentally infected baboons and intrathoracically injected into wild-caught *Chrysops* developed into infective larvae after 14 days of fly rearing with high resultant yields of L3 (Ndzeshang et al. 2020). Validation experiments with these experimentally reared *L. loa* L3 demonstrated they could be cultured to undergo L3-L4 molting *in vitro* and developed into adult stages in C57BL/6 RAG2−/−IL-2gc−/− mice.

4 Conclusion

In the past decade, a significant progress has been made in the development of experimental models for the tropical neglected disease, loiasis. These platforms offer the opportunity to consider how they could be applied in the future to advance filariasis research and development, especially in the areas of host–parasite interaction, drug screening, immune responses and vaccine development, pathogenesis, and diagnosis.

4.1 Perspective of Loiasis Experimental Models for the Advancement of Preclinical Research

Now that tools for basic biology research of loiasis are available, an integral aspect for further investigation is a better understanding of the host–parasitological interactions, for instance, how periodicity is most likely to be influenced by host factors, dissecting stage-specific immune responses to larvae, adults and mf, treatment response to candidate drugs, and scale-up research in vaccine development. New drugs effective in killing adult *Onchocerca* and having little or no effect on *L. loa* mf would be suitable for the control of onchocerciasis, especially in loiasis co-endemic areas. Drug development research on Onchocerciasis focuses mainly on the bacteria Wolbachia that lives in symbiotic relationship with the parasite. However, the scientific assumption that anti-Wolbachia compounds may not have killing properties on *L. loa* is not clearly certain. Thus, the loiasis microfilaremic mouse model serves as an ideal platform to scrutinize for "off-target" rapid, direct microfilaricidal activities of such drugs. Furthermore, the adult loiasis research model is ready to test in vivo treatment responses of candidate macrofilaricide/microfilaricide and for the potential repurposing of other drugs for other needed medical purposes. This infection model could also be readily applied in the preclinical discovery of *L. loa* biomarkers triggered during situations of acute, systemic, drug-induced adverse reaction and evaluation of specific *Onchocerca* candidate biomarkers currently in development as potential point-of-care diagnostics. Additionally, the route course of inflammatory reactions in loiasis microfilaremic mice following treatment with ivermectin can also be considered for pharmacological and biological interventions. Beyond translational research, the murine models will be of benefit to basic parasitological researchers by facilitating a convenient, abundant source of all the mammalian life cycle stages of *L. loa* parasites for molecular and genomic studies as a welcome alternative/refinement to using nonhuman primates.

References

Akue JP, Dubreuil G, Moukana H (2001) The relationship between parasitological status and humoral responses to *Loa loa* antigens in the Mandrillus sphinx model after immunization with irradiated L3 and infection with normal L3. Parasitology 123:71–76

Bain O, Wanji S, Enyong P, Petit G, Noireau F, Eberhard ML, Wahl G (1998) New features on the moults and morphogenesis of the human filarial *Loa loa* by using rodent hosts. Consequences. Parasite 5:37–46

Bain O, Wanji S, Vuong PN, Maréchal P, Le Goff L, Petit G (1994) Larval biology of six filariae of the sub-family Onchocercidae in a vertebrate host. Parasite 1(3):241–254

Cadman ET, Thysse KA, Bearder S, Cheung AYN, Johnston AC, Lee JJ, Lawrence RA (2014) eosinophils are important for protection, immunoregulation and pathology during infection with nematode microfilariae. PLoS Pathog 10(3)

Centre for Disease Control and Prevention (2008) Filariasis. http://www.dpd.cdc.gov/dpdx/HTML/Frames/AF/Filariasis/body_Filariasis_L_loa.htm. Accessed Apr 2023

Chunda VC, Ritter M, Bate A, Gandjui NVT, Esum ME, Fombad FF, Njouendou AJ, Ndongmo PC, Taylor MJ, Hoerauf A, Layland LE, Turner JD, Wanji S (2020) Comparison of immune responses to Loa loa stage-specific antigen extracts in Loa loa-exposed BALB/c mice upon clearance of infection. Parasit Vectors 13(1). https://doi.org/10.1186/s13071-020-3921-x

Desjardins CA, Cerqueira GC, Goldberg JM, Hotopp JCD, Haas BJ, Zucker J, JMC R, Saif S, Levin JZ, Fan L, Zeng Q, Russ C, Wortman JR, Fink DL, Birren BW, Nutman TB (2013) Genomics of *Loa loa*, a Wolbachia-free filarial parasite of humans. Nat Genet 45:495–500

Dieki R, Nsi-Emvo E, Akue JP (2022) The Human Filaria *Loa loa*: Update on Diagnostics and Immune Response. *Research and Reports in Tropical Medicine* 13:41–54

Duke BOL (1955) The development of Loa in flies of the genus Chrysops and the probable significance of the different species in the transmission of loiasis. Trans R Soc Trop Med Hyg 49:115–121

Duke BOL (1960a) Studies on loiasis in monkeys. II. The population dynamics of the microfilariae of Loa in experimentally infected drills (Mandrillus leucophaeus). Ann Trop Med Parasitol 54: 15–31

Duke BOL (1960b) Studies on loiasis in monkeys. III. The pathology of the spleen in drills (Mandrillus leucophaeus) infected with Loa. Ann Trop Med Parasitol 54:141–146

Duke BOL (1964) Studies on loiasis in monkeys. IV. Experimental hybridization of the human and simian strains of Loa. Ann Trop Med Parasitol 58:390–408

Duke BOL (2004) Failed attempts at experimental transplantation and transmission of nocturnally periodic simian Loa from monkey to man. Filaria J 3:S5

Duke BOL, Wijers DJB (1958) Studies on loiasis in monkeys. I. The relationship between human and simian Loa in the rain-forest zone of the British Cameroons. Ann Trop Med Parasitol 52: 158–175

Fombad FF, Njouendou AJ, Ndongmo PC, Ritter M, Chunda VC, Metuge HM, Kwenti TDB, Forsbrook G, Steven A, Cook D, Enyong P, Wanji S, Taylor MJ, Turner JD (2019) Effect of flubendazole on developing stages of Loa loa in vitro and in vivo: a new approach for screening filaricidal agents. Parasit Vectors 12(1). https://doi.org/10.1186/s13071-018-3282-x

Halliday A, Guimaraes AF, Tyrer MHM, Patrick CNW, Arnaud KOJ, Kwenti TDB, Forsbrook G, Steven A, Cook D, Enyong P, Wanji S, Taylor MJ, Turner JD (2014) A murine macrofilaricide pre-clinical screening model for onchocerciasis and lymphatic filariasis. Parasit Vectors:7. https://doi.org/10.1186/s13071-014-0472-z

Hawking F (1955) Periodicity of microfilariae of Loa loa. Trans R Soc Trop Med Hyg 49:132–142. doi:D - CLML: 5528:24727:335 OTO - NLM

Kilarski WW, Martin C, Pisano M, Bain O, Babayan SA, Swartz MA (2019) Inherent biomechanical traits enable infective filariae to disseminate through collecting lymphatic vessels. Nat Commun 10(1). https://doi.org/10.1038/s41467-019-10675-2

Ndzeshang LB, Fombad FF, Njouendou AJ, Chunda VC, Gandjui NVT, Akumtoh DN, Patrick WNC, Andrew S, Pionnier N, Layland EL, Ritter M, Hoerauf A, Taylor JM, Turner JD, Wanji S (2020) Generation of Loa loa infective larvae by experimental infection of the vector, chrysops silacea. PLoS Negl Trop Dis 14(8). https://doi.org/10.1371/journal.pntd.0008415

Orihel TC, Eberhard ML (1985) Loa loa: development and course of patency in experimentally-infected primates. Trop Med Parasitol 36:215–224

Orihel TC, Moore PJ (1975) Loa loa: experimental infection in two species of African primates. Am J Trop Med Hyg 24:606–609

Padgett JJ, Jacobsen KH (2008) Loiasis: African eye worm. Trans R Soc Trop Med Hyg 102:983–989

Petit G, Diagne M, Maréchal P, Owen D, Taylor D, Bain O (1992) Maturation of the filaria Litomosoides sigmodontis in BALB/c mice; comparative susceptibility of nine other inbred strains. Ann Parasitol Hum comparée 67(5). https://doi.org/10.1051/parasite/1992675144

Pinder M, Everaere S, Roelants GE (1994) Loa loa: immunological responses during experimental infections in mandrills (Mandrillus sphinx). Exp Parasitol 79:126–136

Pionnier N, Sjoberg H, Furlong-Silva J, Marriott A, Halliday A, Archer J, Steven A, Taylor MJ, Turner JD (2020) Eosinophil-mediated immune control of adult filarial nematode infection can proceed in the absence of IL-4 receptor signaling. J Immunol 205(3). https://doi.org/10.4049/jimmunol.1901244

Pionnier NP, Sjoberg H, Chunda VC, Fombad FF, Chounna PW, Njouendou AJ, Metuge HM, Ndzeshang BL, Gandjui NV, Akumtoh DN, Tayong DB, Taylor MJ, Wanji S, Turner JD (2019) Mouse models of Loa loa. Nat Commun 10:1429

Sadia A, Fisher M, Juckett G (2008) The African eye worm: a case report and review. J Travel Med 15(1):50–52

Schmidt GD, Roberts LS (2006) Foundations of parasitology, 6th edn. McGraw Hill

Sjoberg HT, Pionnier N, Aljayyoussi G, Metuge HM, Njouendou AJ, Chunda VC, Fombad FF, Tayong DB, Gandjui NVT, Akumtoh DN, Chounna WN, Ndzeshang BL, Lachau S, Tekle F, Quirynen L, Engelen M, Baeten B, Ward SA, Taylor MJ, Wanji S, Turner JD (2019) Short-course, oral flubendazole does not mediate significant efficacy against Onchocerca adult male worms or Brugia microfilariae in murine infection models. PLoS Negl Trop Dis 13:e0006356

Tendongfor N, Wanji S, Ngwa JC, Esum ME, Specht S, Enyong P, Matthaei KI, Hoerauf A (2012) The human parasite Loa loa in cytokine and cytokine receptor gene knock out BALB/c mice: survival, development and localization. Parasit Vectors 5(1). https://doi.org/10.1186/1756-3305-5-43

Turner JD, Pionnier N, Furlong-Silva J, Sjoberg H, Cross S, Halliday A, Guimaraes AF, Cook DAN, Steven A, Rooijen NV, Allen JE, Jenkins SJ, Taylor MJ (2018) Interleukin-4 activated macrophages mediate immunity to filarial helminth infection by sustaining CCR3-dependent eosinophilia. PLoS Pathog 14(3). https://doi.org/10.1371/journal.ppat.1006949

Wahl G, Georges AJ (1995) Current knowledge on the epidemiology, diagnosis, immunology, and treatment of loiasis. Trop Med Parasitol 46(4):287–291

Wanji S, Chunda VC, Fombad FF, Njouendou AJ, Gandjui NVT, Ritter M, Enyong PA, Mackenzie C, Taylor MJ, Hoerauf A, Turner JD (2021) Advances in preclinical platforms of *Loa loa* for filarial neglected tropical disease drug and diagnostics research. Front Trop Dis 2: 778724. https://doi.org/10.3389/fitd.2021.778724

Wanji S, Eyong EE, Tendongfor N, Ngwa C, Esuka E, Kengne-Ouafo AJ, Datchoua FP, Enyong P, Hopkins A, Mackenzie D, Duke B (2015) Parasitological, hematological and biochemical characteristics of a model of hyper-microfilariaemic loiasis (*Loa loa*) in the Baboon (Papio anubis). PLoS Negl Trop Dis 9(11)

Wanji S, Eyong EJ, Tendongfor N, Ngwa CJ, Esuka EN, Kengne-Ouafo AJ, Datchoua-Poutcheu FR, Enyong P, Agnew D, Eversole RR, Hopkins A, Mackenzie CD (2017) Ivermectin treatment of Loa loa hyper-microfilaraemic baboons (Papio anubis): assessment of microfilarial load reduction, haematological and biochemical parameters and histopathological changes following treatment. PLoS Negl Trop Dis 11(7)

Wanji S, Tendongfor N, Vuong PN, Enyong P, Bain O (2002) The migration and localization of *Loa loa* infective and fourth-stage larvae in normal and immunosuppressed rodents. Ann Trop Med Parasitol 96(8). https://doi.org/10.1179/000349802125002220

Whittaker C, Walker M, Pion SDS, Chesnais CB, Boussinesq M, Basáñez MG (2018) The population biology and transmission dynamics of *Loa loa*. Trends Parasitol 34:4

Zofou D, Fombad FF, Gandjui NVT, Njouendou AJ, Kengne-Ouafo AJ, Ndongmo PWC, Datchoua-Poutcheu FR, Enyong PA, Bita DT, Mark J, Taylor DT, Turner JD, Wanji S (2018) Evaluation of in vitro culture systems for the maintenance of microfilariae and infective larvae of Loa loa. Parasit Vectors 11:275

Epidemiology and Public Health Importance

Michel Boussinesq

Abstract

This chapter presents information on the distribution of loiasis and factors explaining variations in prevalence levels. Data on the cases reported from areas outside the classical distribution area and the relationships between human and simian loiasis are also provided. The relationships between entomological, parasitological, clinical, and serological indicators are summarized. Aspects concerning the burden of loiasis in terms of morbidity, mortality, and economic impact, as well as the impact of post-ivermectin serious side effects on onchocerciasis and lymphatic filariasis control programs are discussed.

Keywords

Loa loa · Loiasis · Epidemiology · Humans · Non-human primates · Geographic distribution · Transmission · Population structure · Morbidity · Mortality · Burden

1 Introduction

The relationship between the adult and larval stages (microfilariae) of *Loa loa*, the link between the parasite and some of the manifestations of loiasis (particularly the episodes of transient angioedema called "Calabar swellings"), and the discovery of the vectors of loiasis were elucidated at the end of the nineteenth and beginning of the twentieth century. However, the fact that this filariasis is restricted to central Africa (including Nigeria) and that its manifestations have for long been regarded as

M. Boussinesq (✉)
Institut de Recherche pour le Développement, Unité Mixte Internationale (UMI) 233, Institut National de la Santé et de la Recherche Médicale (INSERM) U1175, Université de Montpellier, Montpellier Cedex, France
e-mail: michel.boussinesq@ird.fr

J. P. Akue (ed.), *Loa loa: Latest Advances in Loiasis Research*,
https://doi.org/10.1007/978-3-031-49450-5_2

mild meant that, before the 1990s, its epidemiology was investigated only by a few research teams, especially in Cameroon and the Republic of Congo (Congo-Brazzaville). Beginning in the 1990s, there was a surge of interest in loiasis because of the occurrence of serious adverse events (SAEs) after ivermectin treatment in individuals harboring high microfilarial densities (MFD) of *L. loa* in the blood—this was a period during which annual mass treatment with ivermectin was implemented to control onchocerciasis in all areas where this disease constituted a public health problem. New studies, conducted to solve this issue, led to major advances in our understanding of the epidemiology of loiasis. However, many unknowns remain, particularly with respect to the importance of the complications of *L. loa* infection. This chapter aims to summarize current knowledge in this domain.

2 Epidemiology

2.1 Geographic Distribution

Reviews published before the 2000s showed that the western, eastern, northern, and southern limits of the transmission area of *L. loa* were located in the south of Benin, the Republic of South Sudan, Southern Chad, and Angola, respectively (Rodhain and Rodhain-Rebourg 1973; Hawking 1977; Boussinesq and Gardon 1997). However, data on endemicity levels within this area were limited, and no information was available for some regions. Beginning in 2001, surveys were conducted in more than 5000 villages to evaluate the prevalence of loiasis using a rapid assessment method called RAPLOA. RAPLOA involves the administration of a questionnaire to a sample of 80 residents aged 15 years and over in the location in question, and assessing the proportion of individuals with a history of subconjunctival migration of an *L. loa* adult worm ("eyeworm"), a pathognomonic sign of loiasis (Takougang et al. 2002). Undertaking these surveys generated a more precise and granular map of prevalence throughout the endemic area (Zouré et al. 2011). Additional data collected in Nigeria and Angola, revealing the presence of isolated foci in the south of the latter country, were subsequently added (Vinkeles Melchers et al. 2020). The resulting map showed clearly that there were two large foci of endemicity, one covering the southern part of Cameroon, the west of the Central African Republic, Gabon, the continental part of Equatorial Guinea, and the northern and western parts of the Republic of Congo, and the other, the northeastern part of the Democratic Republic of Congo (DRC) and some parts of South Sudan (Fig. 1). These two blocks are connected by a transmission zone covering the Sud-Ubangi and Nord-Ubangi Provinces in DRC. On the other hand, the "Cuvette centrale" area, located south of this strip, shows very low levels of endemicity or is non-endemic for loiasis. Low loiasis transmission levels seem to exist on Bioko Island (Priest and Nutman 2017; Yoboue et al. 2022). *L. loa* has never been reported from the islands of São Tomé and Príncipe (Loiseau et al. 2022).

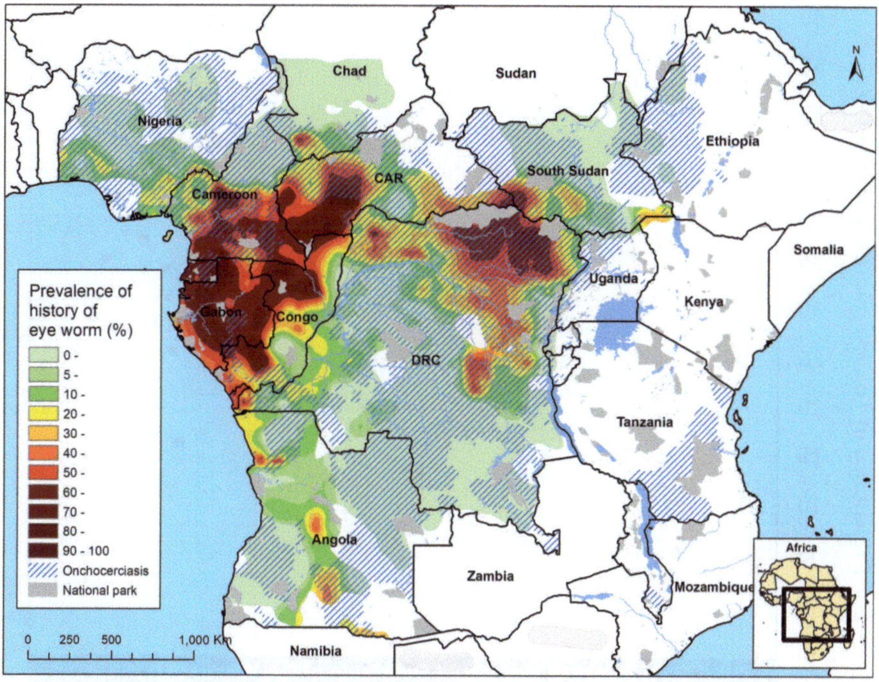

Fig. 1 Map of the estimated pre-control overlap between the prevalence of palpable onchocercal nodules and the prevalence of a history of eye worm in the African Programme for Onchocerciasis Control countries. Abbreviations: *CAR*: Central African Republic; *DRC*: Democratic Republic of Congo. Figure extracted from Vinkeles Melchers et al. (2020). *Clinical Infectious Diseases*, **70**(11): 2281–2289

2.2 Definition of Loiasis Endemicity Levels

Historically, loiasis endemicity levels in a community were defined mainly by the prevalence of microfilaremia (PMF). In the early 2000s, these levels were categorized according to the level of risk for SAEs during mass ivermectin treatment for onchocerciasis control (itself related to the levels of microfilariae in the blood). It has been shown that (1) individuals with a MFD above 8000 microfilariae per milliliter of blood (mfs/mL) have a risk of developing a so-called marked reaction (i.e., with functional impairment lasting several days) after ivermectin treatment (Gardon et al. 1997) and (2) when the PMF in the population aged 15 years and over exceeds 20%, then the proportion of individuals with more than 8000 mfs/mL exceeds 5% (Boussinesq et al. 2001; Takougang et al. 2002). This proportion of 5% was deemed high enough to justify the implementation of strengthened surveillance procedures in areas where the PMF exceeded 20% (Mectizan Expert Committee and Technical Consultative Committee 2004). As the RAPLOA studies had shown that this PMF typically corresponded to a proportion of people with a history of eyeworm (PEW) of 40% (Wanji 2001; Takougang et al. 2002; see Fig. 2), the

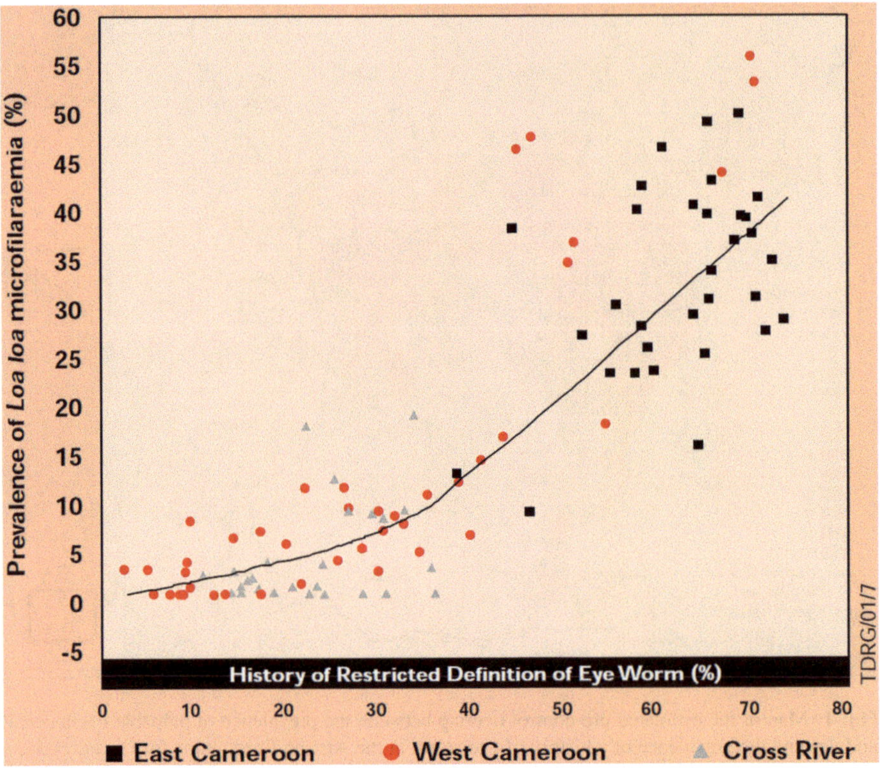

Fig. 2 Relationship between prevalence of microfilaremia and Rapid Assessment Procedure based on the Restricted Definition of Eye Worm. Figure extracted from Wanji (2001). TDR/IDE/RP/RAPL/01.1

latter value was subsequently adopted to define at-risk areas (Mectizan Expert Committee and Technical Consultative Committee 2004) and communities with a PEW ≥40% were labeled "high-risk" or "hyperendemic" (Zouré et al. 2011). In other studies, a distinction was made between communities with a PEW <20% ("low loiasis"), a PEW from 20% to 40% ("medium loiasis"), and a PEW >40% ("high loiasis") (Kelly-Hope et al. 2014).

2.3 Factors Explaining the Distribution of Loiasis and the Variations in Loiasis Endemicity Levels

The distribution of loiasis is governed by the distribution of its vectors, which are tabanids belonging to the genus *Chrysops*. The two main species transmitting *L. loa* to humans are *C. silacea and C. dimidiata*. Their distribution has been reviewed by various authors including Oldroyd (1957), Crosskey et al. (1980), and Kelly-Hope et al. (2017). These species are present in forested or forest mosaic areas between

Nigeria (on the west) and South Sudan (on the east); they have also been reported in Ghana (Austen 1909; Crosskey et al. 1980) but their presence in this country has not been confirmed recently. The two species bite during the day (often with a peak in the morning and another in the afternoon) and are highly anthropophilic—a study in the Republic of Congo showed that 80–90% of their blood meals had been taken on humans (Gouteux et al. 1989). Their presence and population density depend on well-known environmental factors (Badia-Rius et al. 2019). The adult flies typically rest in the canopy and are attracted to the ground by visual stimuli (some specific colors, particularly light blue and movement, particularly of groups of people) but also, particularly for *C. silacea*, by the smell of wood fires (Duke 1955a; Caubère and Noireau 1991). They avoid bright sunlight or areas of dense shade and prefer light shade and forest fringe areas. Cocoa, rubber, and palm tree plantations are favorable biotopes for *C. silacea* and *C. dimidiata*. They can reach a maximum range of 4.5 km within 2 weeks (Chippaux et al. 2000). Both species are usually found in sympatry, with one species more abundant than the other, depending on the season (Davey and O'Rourke 1951; Duke 1959;
et al. 1990b, 1991). Previous work suggests that the presence of *C. dimidiata* is typically more confined to forested zones than *C. silacea*, the adults of the latter species being more able to reach open areas such as villages (Kelly-Hope et al. 2017). The oviposition places of these species are reasonably well-characterized. Adult females lay their eggs on the leaves of plants growing on mud covered by decaying vegetation and a film of slow-moving water (Crewe and Williams 1961). Environmental factors governing the distribution and density of the vectors include the type of vegetation favorable to the adult stages and the type of soil favorable to the development of the larvae, as well as temperature, humidity, and altitude characteristics. Notably, these factors explain the absence of these species in the "Cuvette centrale" area (Kelly-Hope et al. 2012). Geostatistical models have been developed to predict the prevalence of loiasis accounting for several environmental parameters, including altitude and the Normalized Difference Vegetation Index (NDVI) (Thomson et al. 2004; Diggle et al. 2007; Schlütter et al. 2016).

In addition to the species highlighted above, there are several other species of *Chrysops* that might play a minor role in *L. loa* transmission, though many of these have a crepuscular peak biting time and are much less anthropophilic than *C. silacea* and *C. dimidiata,* and development of the parasite is slower. Duke (1955b) suggests that *C. zahrai*—a species reported from northwest Cameroon at the fringe between mountain forest (~1500 m) and grassland, as well as from Nigeria (Crosskey et al. 1980)—could be a subsidiary vector of some local importance in areas where the prevalence of infection is high in the human population. Duke (1955b) also thinks that *C. centurionis*, a forest species present from Nigeria to Uganda, could bite humans during evening forest activities. Woodman and Bokhari (1941) hypothesize that *C. distinctipennis* (a savanna species present from Mauritania to Uganda) and *C. longicornis* (widely present in Africa, both in savanna and in forest) could be the vectors of *L. loa* in South Sudan, even if their vectorial competence is much lower than that of *C. silacea* and *C. dimidiata*. These two species are also present in Chad (Taufflieb and Finelle 1956) and could maintain the loiasis focus located in the south of the country, where the prevalence of infection is surprisingly high for a savanna area. *C. dimidiata* and *C. longicornis* are present in Northern Angola (Casaca 1966), but the *Chrysops* species present in the southern part of this country are unknown.

2.4 Sporadic Cases of Human Loiasis Outside the Classical Distribution Area

Outside the classical endemic range of loiasis described above, cases of loiasis have been reported in people from several West African countries (Burkina Faso, Ghana, Republic of Guinea/Guinée-Conakry, Liberia, Mali, and Sierra Leone) who had never traveled to areas east of Benin (Thorpe 1892; Léger 1912; Corson and Ingram 1923; Blacklock 1930; Strong et al. 1930; Willcox 1945; Hughes and Sarkies, 1951; Poindexter 1953; Wright 1961; Volpicelli et al. 2020; Nicolini et al. 2022). Cases have also been reported from western Uganda (Price 1961; Nnochiri 1972; Poltera 1973), the southwestern part of Ethiopia (White 1977), and Zambia (Buckley 1946; Chhabra et al. 1989). It should be noted that most of these cases were reported more than 60 years ago. It is therefore difficult to say whether these reports are due to erroneous diagnosis (e.g., *Wuchereria bancrofti* and *L. loa* microfilariae (mfs) are of similar size, and the former, whose density is generally higher at night, can also be found in day blood in cases of high MFD), whether the collected travel history for each report was incomplete (omitting some travel to an endemic area), or whether the subjects had indeed been infected locally by a *Loa* parasite of simian origin transmitted by *C. distinctipennis* or *C. longicornis*—baboons naturally infected with *Loa* have been reported from the Republic of Guinea (Treadgold 1920) and Uganda (Nelson 1965) (see below). In the case of the reports from Uganda, some appeared as nodules in the bulbar conjunctiva (so-called Kampala eyeworm) (Nnochiri 1972; Poltera 1973), and authors have suggested that these observations could actually be due to infections with *Mansonella* sp., possibly *M. perstans* (Baird et al. 1988; Orihel and Eberhard 1998). It is interesting to note that the parasites collected from the two recent cases from Republic of Guinea (Volpicelli et al. 2020; Nicolini et al. 2022) have been confirmed to be *L. loa* by molecular techniques.

Cases of infections with *L. loa* have also been reported in Asia (mainly India) from subjects who had apparently never visited a classical endemic area (e.g., Mandal et al. 2013; Kumari et al. 2019), though the information provided in the publications is not sufficient to confirm the legitimacy of the species diagnosis. Indeed, in some cases, the information shows that the diagnosis is clearly erroneous. However, it should be noted that a parasite named provisionally *Loa inquirenda*, whose adult stage is 2–3 times longer than the African *Loa* but has similar morphologic characteristics (including cuticular bosses), has previously been described in India (Maplestone 1938).

L. loa does not seem to have established on the American continent, nor in the Caribbean islands, during the Atlantic slave trade, despite the fact that *L. loa* can develop—in experimental conditions—up to the stage of infective larvae in *Chrysops atlanticus*, a species present all along the East Coast of the United States

(Orihel and Lowrie 1975). Flies of the genus *Chrysops* are present in all continents (https://www.gbif.org/fr/species/1497045). However, other than with *C. atlanticus*, no attempt has been made to evaluate whether species present outside Africa can transmit *L. loa*.

Several studies have been conducted to determine whether arthropods other than *Chrysops* species can transmit *L. loa*: These include tabanids belonging to the genera *Haematopota* (four species tested), *Tabanus* (four species) and *Hippocentrum* (*H. trimaculatum*), stomoxes (two species), tsetse flies (two species), mosquitoes (*Anopheles maculipennis*, *Aedes aegypti*, *Culex pipiens*, and *Mansonia africana*), fleas (*Pulex irritans*), bedbugs (*Cimex rotundatus*), and ticks (*Ornithodoros moubata*) (Fülleborn 1908; Leiper 1913; Woodman and Bokhari 1941; Ogunba 1972). Partial development of the parasite was reported in *Haematopota* and in *H. trimaculatum* (Leiper 1913; Woodman and Bokhari 1941), and complete development up to the stage of infective larvae was found in 5.1% of the *M. africana* fed experimentally (Ogunba 1972). This species is widely distributed in Africa (Moraga et al. 2015). No development of *L. loa* was found in the other arthropods tested.

2.5 Human and Simian Loiasis

Loa sp. can also be found in non-human primates (NHP). Natural infections with *Loa* have been reported from several species belonging to the Cercopithecidae family: These include *Papio papio* baboons captured in the Republic of Guinea (Treadgold 1920) (adult worms were found in 13 of the 55 autopsied baboons); one grey-cheeked mangabey (*Lophocebus* [formerly *Cercocebus*] *albigena johnstoni*) captured in the Bas-Uélé region in DRC (Sandground 1936); drills (*Mandrillus leucophaeus*) and three *Cercopithecus* species (*C. nictitans martini*, *C. mona mona*, and *C. preussi*) killed in the South-West region of Cameroon (Duke and Wijers 1958), and *Papio doguera* (now *P. anubis*) baboons and *Cercopithecus aethiops*, *C. mitis*, *C. nictitans*, and *Colobus abyssinicus* in the northwestern part of Uganda, in the Budongo forest (Nelson 1965). Adult stages of *Loa* have also been found in mountain gorillas (*Gorilla gorilla beringei*) in the Kivu Province (DRC), and blood mfs were identified in 4 of the 20 animals examined (Van den Berghe et al. 1957, 1964). One western lowland gorilla (*Gorilla gorilla gorilla*) was also found harboring adult worms in Gabon (Bain et al. 1995). One eastern chimpanzee (*Pan troglodytes schweinfurthii*) captured in DRC was also found infected with *Loa*—two adults were recovered from the animal (Rodhain and Van den Berghe 1939; Rodhain 1946). Lastly, one South American NHP (*Ateles paniscus*) was found to be infected in the Hamburg Zoo (Vogel 1927), and Nelson (1965) suggested that this animal had perhaps been infected during its transit in West Africa or during its captivity in Europe, where several species of *Chrysops* are found. It should be noted that most of the studies on natural infections of NHPs by *L. loa* were published more than 60 years ago.

Experimental infection of NHPs with human *L. loa* was successful with several species belonging to the Cercopithecidae family: This includes drills, mandrills

(*Mandrillus sphinx*), baboons (*Papio anubis*), patas monkeys (*Erythrocebus patas*), and rhesus monkeys (*Macaca mulatta*, which lives in Asia) (e.g., Duke 1957; Eberhardt and Orihel 1981; Orihel and Eberhardt 1985; Dennis et al. 1993; Pinder et al. 1994; Wanji et al. 2015a; Akué et al. 2016). Attempts failed in green monkeys (*Cercopithecus aethiops*, n = 2), one chimpanzee (*Pan troglodytes*) (Orihel and Moore 1975), and crab-eating macaques (*Macaca fascicularis*, n = 5) (Pinder et al. 1990), which live in Southeast Asia.

Dissection of a significant number of NHPs performed by Duke and Wijers (1958) in Cameroon showed that 96% of the drills harbored *Loa* adult worms and that infection rates were much lower in the two *Cercopithecus* species studied (24% for *C. nictitans* and 12% for *C. mona*). This study also showed that the *Loa* adult stages collected in NHPs were significantly longer than those obtained from humans and that, in NHPs, *Loa* mfs generally showed nocturnal periodicity in the peripheral blood. This contrasts with the diurnal periodicity observed in humans. However, *Loa* of human and simian origins do belong to the same species because fertile hybrids can be obtained (Duke 1964). The main vectors of simian *Loa* are *C. langi* and *C. centurionis,* which have a crepuscular peak of activity (between 5 pm and 9 pm), live in the forest canopy, and do not usually bite humans (Duke 1957). Similarly, the analysis of 408 blood meals taken by *C. dimidiata* and *C. silacea* caught in the Chaillu Mountains in the Republic of Congo did not identify blood meals that had been taken on NHPs (Noireau and Gouteux 1989). Given the differences in mf periodicity (diurnal versus nocturnal), timing of peak vector activity (diurnal versus crepuscular), and host preference of the *Chrysops* species biting humans and NHPs, most authors consider the two transmission cycles (in humans and NHPs) to be separate, and conclude that NHPs do not constitute a reservoir of *Loa* for humans (Duke and Wijers 1958; Nelson 1965; Rodhain 1980; Noireau and Gouteux 1989) and that the two "strains" of *Loa* are in the process of speciation (Fain 1978). However, as mentioned above, exceptional cases of human infection by *Loa* parasites of simian origin cannot be definitively excluded. Such cases could explain the observation that, in some loiasis foci, some microfilaremic individuals have similar or even higher levels of mf in the blood collected at night as compared to blood collected during the day. Such a phenomenon has been reported from South Sudan (Woodman and Bokhari 1941), Nigeria (Ogunba 1977), and two provinces of DRC: Kongo Central (formerly Bas-Congo) (Fain et al. 1974) and Haut-Uélé (Bakajika et al. 2014).

2.6 Number of Exposed and Infected Individuals

Estimates of the number of individuals infected with *L. loa* have varied considerably over time. In 1988, this number ranged between 2 and 13 million (Pinder 1988). This estimation is complicated by the fact that a high proportion of individuals, who are certainly infected with the parasite (e.g., have a history of eyeworm or Calabar swellings) never have detectable blood mfs (Fain 1978; Pinder 1988) but have what is called "occult loiasis" (Touré et al. 1998). For example, in very highly infected villages of the Lékié division in Cameroon, where the *L. loa* PMF exceeded 35% in the total population aged 5 years and older, the PMF was only 53.5% and 45.6% in males and females aged ≥60 years, a subpopulation where >90% of the individuals were likely infected (Pion et al. 2004). This absence of mfs circulating in the peripheral blood is probably due to immunologic phenomena (Baize et al. 1997; Akué et al. 1997, 1998) and shows a familial genetic predisposition. It has been estimated that 41% of the individuals exposed to the parasite have a very low lifetime risk of becoming microfilaremic (Garcia et al. 1999).

Using the data collected during the RAPLOA surveys and population data from areas where loiasis is endemic, Zouré et al. (2011) estimated that 14.4 million people lived in areas where the PEW exceeded 40% (i.e., where the risk of post-ivermectin SAEs was high). Using the same data, Vinkeles-Melchers et al. (2020) estimated that the numbers of microfilaremic individuals in 1995 and 2015 were 3.7 and 4.0, respectively, and were projected to increase to 6.4 million by 2025.

The total population of all endemic countries is approximately 450 million people. Considering that travel within countries is increasingly frequent and that infection can occur after very short exposure (see below), the number of people exposed to *L. loa* is certainly considerable.

The number of people from non-endemic countries who are infected with *L. loa* during short or long stays in endemic areas is unknown. According to published data, it is likely in the order of a few hundred over 20 years (Lipner et al. 2007; Develoux et al. 2017; Puente et al. 2020; Bottieau et al. 2022), although this is likely an underestimate given that diagnostic expertise is limited in many non-endemic settings.

2.7 Distribution of the Parasites in the Population; Variation in PMF and MFD According to the Population Categories

In endemic areas, the PMF is generally higher in males than in females. It increases rapidly up to the age of 20–30 years and then increases more slowly or plateaus thereafter; in some cases, the PMF decreases somewhat in older age groups (e.g., Kershaw et al. 1953; Ripert et al. 1977; Gryseels et al. 1985; Noireau et al. 1989; Wanji et al. 2003; Pion et al. 2004; Whittaker et al. 2018). The profile of PMF with age seems to vary not only according to gender, but also according to the level of endemicity in the community and to biogeographic factors (Wanji et al. 2003; Pion et al. 2004). MFD differences according to gender and age show a similar profile to that of PMF.

As expected, the risk of *L. loa* microfilaremia is closely associated with regular visits to places exposing people to *Chrysops* bites (Garcia et al. 1999; Pion et al. 2005; Mischlinger et al. 2018). Individuals visiting forested environments are particularly at risk.

To the best of our knowledge, only one study has been conducted comparing the prevalence and intensity of microfilaremia in individuals living in the same area but belonging to different ethnic groups. Noireau et al. (1989) found that the PMF was lower in Pygmies than in Bantu people living in the same village in the Republic of

Congo, even though the frequency of eyeworm episodes was similar in the two groups, suggesting that they harbored, on average, a similar number of adult worms.

The distribution of MFDs in the human population is very overdispersed. In villages of central Cameroon, it could be described by a zero-truncated negative binomial distribution, and the aggregation parameter k seemed to be similar irrespective of the age of the host or the level of endemicity in the community. This parameter k for *L. loa* was around 0.3, a value close to those previously reported for *Wuchereria bancrofti* and *Onchocerca volvulus* (Pion et al. 2006b). A Weibull distribution has also been proposed to describe the *L. loa* MFD distribution (Schlüter et al. 2016).

Some individuals have very high MFDs, exceeding 100,000 mfs/mL (Chippaux et al. 1996; Boussinesq et al. 1998). As for occult loiasis, studies have indicated that some individuals have a familial predisposition to high levels of microfilaremia (Eyebe et al. 2018). This phenomenon can lead to geographical clustering of very highly microfilaremic individuals in some villages. In turn, this clustering explains (in part) the weak correlation observed between the PMF and the proportion of individuals with >30,000 mfs/mL (Boussinesq et al. 2001; Wanji 2001).

As reported in the 1990s, MFDs can remain stable at an individual level for several months or years (2–3 years) (Noireau and Pichon 1992; Garcia et al. 1995). A more recent study comparing individual MFDs at 5-day, 1-month, and 16-month intervals showed that this stability is relative for some individuals (Campillo et al. 2023a).

When considering only microfilaremic individuals in a community, the mean MFD seems to be relatively constant with age, including in younger age groups (Garcia et al. 1995; Pion et al. 2004). This observation could suggest that individuals pass from an amicrofilaremic status to a steady-state MFD within a fairly short interval (possibly on the order of several weeks or months), and that this subsequently remains relatively stable for years.

2.8 Transmission, Vector-Parasite Relationship, Seasonality

2.8.1 Relationships Between MFD, Number of mfs Ingested by the Vector, and Number of Infective Larvae

The number of mfs ingested by *Chrysops* which have taken a bloodmeal on a given individual can vary widely, and this number is generally lower than what would be expected by combining the subject's MFD and the volume of the bloodmeal (Kershaw et al. 1954). This is probably because *Chrysops* feed from a pool of blood and not directly from a capillary (Kershaw et al. 1954).

Kershaw and Duke (1954) concluded that most, if not all, of the *L. loa* mfs ingested by a *Chrysops* reach the stage of infective larvae. However, reanalysis of the data suggests that only 40–50% of the parasites progress to this stage (Whittaker et al. 2018). It is not known whether this proportion varies in a density-dependent manner with microfilarial intake.

The number of infective larvae found in naturally infected *Chrysops* can be extremely high. In a study conducted in southwest Cameroon, the mean number of infective larvae in the heads (L3H) of infective *C. silacea* was 146.1, and some flies harbored 600 L3H. In this study, the distribution of the number of infective larvae could be described by a negative binomial distribution (Wanji et al. 2002). However, in other studies of *Chrysops* natural infection levels, the mean number of L3H per infective *Chrysops* was much lower: Considering *C. dimidiata* and *C. silacea* separately, this number was 10.1 and 11.2, respectively, in a study conducted in the Republic of Congo (Noireau et al. 1990b) and 36.5 and 81.5 in Kokodo, a village located in the Center Region of Cameroon (Demanou et al. 2001). This mean number was 9.8 in southeast Nigeria for both species combined (Uttah 2013).

2.8.2 Seasonal Fluctuations in *Chrysops* Abundance and Transmission Potentials

Seasonal fluctuations in *Chrysops* population density have only been investigated in a limited number of studies. This is probably because loiasis is generally not regarded as a public health problem, *Chrysops* densities are usually low, and it is more difficult to capture *Chrysops* than other vectors (mosquitoes, blackflies, etc.). These studies were conducted in south Nigeria (Davey and O'Rourke 1951; Uttah 2013), in southwest Cameroon (Duke 1959), in the Lekoumou division of the Republic of Congo (Noireau et al. 1990b, 1991), and in the Lékié division in Cameroon (Demanou et al. 2001). They showed that *Chrysops* densities vary widely according to the rainfall and the temperature, sometimes with marked differences between C. *dimidiata* and *C. silacea*. In general, however, results suggest that rainfall induces the emergence of adult *Chrysops*, with densities usually higher during the hot rainy season and lower during the dry season.

L. loa transmission potentials also fluctuate throughout the year but this is more difficult to characterize than fluctuations in abundance (Duke 1959; Noireau et al. 1990b; Demanou et al. 2001).

2.9 Relationships Between Entomological, Parasitological, Clinical, and Serological Indicators

2.9.1 Relationship Between Entomological and Parasitological Indicators

Very few studies have been conducted characterizing the relationship between entomological (particularly the Annual Transmission Potential, ATP) and parasitological (prevalence) indicators of loiasis. A study performed in four villages of the Republic of Congo showed that, to a certain extent, there was a relationship between the *Chrysops* biting rate (evaluated over 3 months only) and the PMF (Noireau et al. 1991). In another study conducted in three Gabonese villages, where the ATPs were 250, 1800, and 43,000 L3s per human per year, the PMFs (assessed by leucoconcentration of 1 mL blood) in the total population (aged 6–80 years) were very similar (21%, 22%, and 22%, respectively). In contrast, the PMFs in children of these villages were 39%, 62%, and 96%, respectively, and the *L. loa* PCR positivity rates were 57%, 84%, and 85%, respectively (Touré et al. 1999). Another study in two of these villages showed that, even though the PMFs were similar, the mean MFD was much higher in the high-transmission village (Akué et al. 2002).

It has been shown that exposure to high *L. loa* transmission levels leads to depressed specific and nonspecific T-cell proliferative responses in infected humans (Akué and Devaney 2002). Transmission intensity also impacts the levels of the various IgG subclasses, with higher levels of L3-specific IgG3 in a low-transmission village and higher levels of mf-specific IgG2 in a high-transmission village (Akué et al. 2002). These differences in the immune responses according to the transmission level might limit the increase in the PMF beyond a given transmission intensity.

2.9.2 Relationship Between PMF and Mean MFD

The relationships between PMF and various quantitative (arithmetic or geometric mean MFD) or semi-quantitative indicators (e.g., proportion of subjects with >8000 or >30,000 mfs/mL) were investigated in several regions of Cameroon, in southeast Nigeria, in the western and northeastern parts of DRC and in Gabon (Boussinesq et al. 2001; Takougang et al. 2002; Akue et al. 2011; Wanji et al. 2012; Schlüter et al. 2016). Although small differences between regression lines were found between sites, they were not considered significant enough to change the conclusion (which has operational implications) that a PMF of 20% corresponds to a ~5% proportion of subjects with >8000 mfs/mL (Wanji et al. 2012).

The proportion of individuals with >30,000 mfs/mL is less closely related to PMF than the proportion of subjects with >8000 mfs/mL, with wider variations in the former indicator for a given PMF value. This is probably because only a few people show such hypermicrofilaremia (~1% when the PMF is 20%) and that, as mentioned above, there is a genetic predisposition to the development of such a high MFD (Eyebe et al. 2018). Thus, if several predisposed families live in a same village, the proportion of hypermicrofilaremic individuals can be high.

2.9.3 Relationships Between Parasitological and Clinical Indicators

Initial surveys conducted to develop and validate the RAPLOA method, as well as several subsequent studies, have facilitated characterization of the relationship between PMF and PEW (Wanji 2001; Takougang et al. 2002; Akue et al. 2011; Wanji et al. 2012; Emukah et al. 2018). During the first surveys, when the PMF in subjects aged >15 years was 5%, 10%, 20%, 30%, or 40%, the PEW in the same subjects was, on average, 23%, 36%, 49%, 62%, and 73%, respectively (Wanji 2001). For PMF exceeding 20%, the PEW might be slightly higher in DRC (Wanji et al. 2012). During a study conducted in 2016 in south Nigeria in districts never treated with ivermectin, the relationship between PMF and PEW was not very different from that described previously but, for a given PEW, the proportion of subjects with very high MFDs (defined as >20,000 mfs/mL in this study) was much lower than that reported during the first RAPLOA surveys (Emukah et al. 2018). This phenomenon could be because this study was performed in districts located near areas treated with ivermectin for many years.

The first RAPLOA surveys also enabled evaluation of the relationship between the PMF and the proportion of individuals with a history of Calabar swellings (PCS). This relationship was found to be much less closely correlated than that between PMF and PEW: When the PMF was 20–50%, the PCS could vary between 40% and 80%. In addition, when the PMF was low, the PCS was on average higher in southeast Nigeria than in southwest and northwest Cameroon (Wanji 2001). For these reasons, questions about a history of Calabar swellings were excluded from the RAPLOA procedure.

2.9.4 Relationship Between Parasitological and Serological Indicators

A rapid test for detection of *L. loa*-specific antibodies has been recently developed (Pedram et al. 2017). This test could be most useful not only for diagnosis of occult loiasis, but also to measure the levels of loiasis endemicity and evaluate the risk of post-ivermectin SAEs. A study was conducted in 146 Gabonese villages to evaluate the relationship between the proportion of subjects with >20,000 mfs/mL and both the seropositivity rate and the PMF (Johnson et al. 2022). Whereas the data showed a clear correlation between parasitological and serological indicators, it also highlighted the insufficiency of seroprevalence alone in classifying villages at risk or without risk of post-ivermectin SAEs. Thus, in some villages, additional parasitological examinations are needed to evaluate and correctly classify the risk level.

2.10 Infections After Very Short Exposure to Chrysops Bites and Clinical and Parasitological Prepatent Periods

A long stay in an area of *L. loa* transmission is not required for successful infection by the parasite. Exposure for several days can be enough, as demonstrated by anecdotal case reports (Richardson et al. 2012; Jazuli et al. 2016).

The duration of clinical prepatence (the interval between an infective bite and the appearance of the first signs and symptoms) is highly variable. Based on studies of expatriates exposed for a fairly short period of time, it can range from 2 months (Sharp 1929) to more than 15 years (e.g., Thomas et al. 1970; Richardson et al. 2012). Among 49 expatriates examined by Churchill et al. (1996), the median interval between the first possible exposure and the appearance of signs was 15 months (range: 5–156 months). Among 47 expatriates studied by Antinori et al. (2012), the median interval between the last possible exposure and the onset of signs was 108 weeks (25.2 months) (range: 2–576 weeks; 0.5–134.4 months).

Very little information exists concerning the duration of parasitological prepatence (interval between infective bite and appearance of blood mfs) in humans. Work by Janssens et al. (1958) suggests it is in the order of at least 17 months, though Churchill et al. (1996) suggest only 6–12 months. Data on PMF in children under 5 years living in endemic villages are also quite sparse. PMF was 2.8% among children aged 1–5 years in the small town of Mbandjok (Cameroon) (Cot et al. 1995), and among the 52 children under 5 years examined in Ngat, a highly endemic village of central Cameroon, only 1 (1.9%) had *L. loa* mfs on a 30-μL-thick film

(Mommers et al. 1995). In splenectomized NHPs experimentally infected with *L. loa*, the period of parasitological prepatence is on average 150 days (5 months), regardless of the species (Boussinesq 2006; Wanji et al. 2015a).

2.11 Coinfections with Other Pathogens

The simultaneous presence of *L. loa* and *M. perstans* mfs in the peripheral blood seems to occur more frequently than would be expected by chance (Noireau et al. 1989). In addition, there appears to be a correlation between the MFDs of these two filarial species (Drame et al. 2016). Infection with the parasite called *Mansonella* sp. "DEUX" is also associated with an increased risk of being infected with *L. loa* and/or *M. perstans* (Sandri et al. 2021). The joint presence of mfs of these last two species in the blood is associated with high titers of total IgE (Bouyou-Akotet et al. 2014). A study conducted in Cameroon also showed that there was a weak positive association between *L. loa* MFD and *O. volvulus* MFD, the former increasing by 11% when the latter increases by 100 mfs per skin snip (Pion et al. 2006a).

There appears to be no association between infection with *L. loa* and *Plasmodium* (Drame et al. 2016; Abbate et al. 2020; Yoboue et al. 2022). *L. loa* infection also does not seem to influence trypanosomiasis serological status assessed by CATT (Pion et al. 2017).

The prevalence of *L. loa* infection appears to be similar in people with and without HIV infection (Pongui Ngondza et al. 2022). However, in subjects infected with HIV, the presence of a *L. loa* microfilaremia is associated with a lower CD4 density and a higher viral load than in amicrofilaremic individuals (Njambe Priso et al. 2018). Although no link has been reported previously (Delaporte et al. 1989), a recent study conducted in Gabon showed a correlation between *L. loa* MFD and HTLV-1 viral load (Akué et al. 2020).

To our knowledge, no study has been conducted to assess the possible interactions between *L. loa* and soil-transmitted helminths and the respective role of these parasites in the occurrence of hypereosinophilia.

3 Public Health Importance of Loiasis

3.1 The Various Types of Manifestations of Loiasis

Loiasis is often regarded as a disease with a low impact in terms of public health because its endemic range is limited to central Africa and because its most common manifestations (pruritus, Calabar swellings, eyeworm) are transient and considered benign. Until recently, only a few authors have mentioned that these manifestations can lead to significant disability and psychological distress (e.g., Johnstone 1947; Gordon et al. 1950; Fain 1978; Marriott 1986). This distress can be all the more marked as loiasis is a chronic disease: The mean lifespan of the adult worm is unknown but it can live up to 20 years (Richardson et al. 2012) and people living in endemic areas are constantly reinfected. Furthermore, it is recognized that loiasis can lead to renal, pulmonary, cardiac, neurological, or splenic complications (Buell et al. 2019).

3.2 Frequency and Severity of the So-Called Benign Manifestations of Loiasis

We reported earlier that, at a community level, the incidence of eyeworm episodes and that of Calabar swellings correlated quite well with the PMF and that the proportion of the population with a history of such signs could be very high. However, the impact of these manifestations in terms of public health largely depends on the duration and frequency of the episodes and the severity of the associated discomfort. These factors remain poorly understood.

It is often thought that the subconjunctival migration of adult *Loa* is rapid (a few minutes to an hour) and therefore that it is necessary to intervene quickly if one wants to capture the worm during the episode. However, an analysis of data collected in eastern Cameroon (Takougang et al. 2007) shows that, in the study population, the proportion of subjects for whom the duration of the episode was <24 h, 24–48 h, 48–72 h, and >72 h was 9%, 25%, 29%, and 37%, respectively (Chesnais unpublished). These results are consistent with those of a study conducted in Gabon, where the median duration of these episodes was 3 days (interquartile range: 1–4 days) (Veletzky et al. 2020). Moreover, in Gabon, these episodes were accompanied by local pain, visual discomfort, and impairment in everyday life and at work in 94%, 79%, and 69% of cases, respectively. In this same study, the median number of eyeworm episodes in the previous year was 1, but some subjects had had 12 (approximately one per month) (Veletzky et al. 2020). Among subjects aged 15 years and over examined in East Cameroon who had an episode of eyeworm in the last 12 months, the proportion of those who had 1, 2, 3, and more than 3 episodes was 21%, 32%, 26%, and 21%, respectively (Chesnais unpublished).

The duration of Calabar swellings is usually considered to last from a few hours to a few days. Among subjects who reported this sign during the study in eastern Cameroon, the proportion of those for whom they had lasted <24 h, 1–2 days, 3–7 days, and >7 days was 7%, 42%, 43%, and 8%, respectively. About half (49%) had had one or two episodes in the past 12 months, and 51% three or more episodes (Chesnais unpublished). In addition, these swellings were associated with local pruritus in more than 90% of cases, and 28% and 46% of subjects declared that they were painful or very painful, respectively (Takougang et al. 2007). This associated pain was also found in the study conducted in Gabon (Veletzky et al. 2020). These swellings can also lead to marked functional impairment (Carme et al. 1989), especially when they limit the mobility of a joint (wrist, etc.).

In addition to these well-known manifestations, loiasis can induce nonspecific symptoms but very few studies have been conducted to assess their frequency. Data collected in the Republic of Congo suggest that infection with *L. loa* is frequently associated with pruritus, most often presenting as sudden episodes, affecting the upper limbs in more than 90% of cases and described as severe in 51% of cases. It is also associated with rash, headache, and arthralgia (Pinder 1988; Carme et al. 1989; Noireau et al. 1990a). In Gabon, loiasis was also significantly associated with paresthesia, fatigue, and headache, which could be severe, leading to impairment of activities of daily living and ability to work (Veletzky et al. 2020).

3.3 Health-Seeking Behavior of Individuals with "Benign" Manifestations of Loiasis

The classic manifestations of loiasis can cause great psychological suffering in affected people (Ndzana 2011). Several studies have reported individuals applying various products to the eye when an adult *Loa* migrates under the conjunctiva. These include products extracted from various plants (Ndzana 2011), including both food plants, such as onion, lemon, garlic or chili (Takougang et al. 2007; Veletzky et al. 2021), and medicinal plants (Mengome et al. 2018). Another method used by infected individuals is to extract the adult worm by incising the conjunctiva with a razor blade or a palm thorn (Takougang et al. 2007). These actions can, of course, cause ocular damage. Similarly, several traditional preparations are used to alleviate Calabar swellings and pruritus (Mengome et al. 2018)—for the latter, self-medication with antihistamines is also common. The choice of traditional techniques or self-medication (using street vendors) is frequently linked to financial and logistical constraints (lack of a nearby health center) but also, sometimes, to the fact that "modern" treatments are considered ineffective (Ndzana 2011).

When the obstacles to access to conventional medicine are removed, loiasis can be a very frequent reason for consultation in health facilities. Very few data exist on the subject but, in the 1980s, loiasis was the third most frequent cause of consultation at the Mossendjo Hospital in the Republic of Congo, after pulmonary diseases and malaria (Boulesteix and Carme 1986). In the same country, during monthly visits carried out in 1982 in three villages in the Mayombe forest, pruritus probably linked to *L. loa* (according to the authors) was the third most frequent cause of consultation by adults, after rheumatic diseases and "asthenia and various subjective disorders" (Richard et al. 1988).

3.4 Frequency of Severe Manifestations of Loiasis

Renal, neurological, cardiac, pulmonary, severe ocular (penetration of the adult worm inside the eyeball and retinal hemorrhages), and splenic complications of loiasis have been described in case reports (Boussinesq, 2006; Buell et al. 2019). They are often considered rare, but in reality, apart from a study of 297 subjects that was limited to cardiac disorders (Tenaguem et al. 2010), no population study has been conducted before 2022 to assess their frequency. Their importance, in terms of public health and overall disease burden, is therefore unknown. Recent studies demonstrated that *L. loa* microfilarial density is associated with proteinuria levels (Campillo et al. 2023b) and with altered cognition (Checkouri et al. 2023).

3.5 Excess Mortality Associated with Loiasis

Two studies were conducted to assess whether loiasis had an impact on the life expectancy of infected subjects. These studies were cohort studies using individual parasitological data (MFD) collected during the development and validation of the RAPLOA method. The first, conducted in 2016 in 28 villages in the East region of Cameroon (and which had been initially examined in 2001), showed that the standardized mortality rate in the villages was significantly correlated with the proportion of subjects harboring more than 30,000 mfs/mL but not to the prevalence of microfilaremia nor to the proportion of subjects with a history of eyeworm or Calabar swelling (Chesnais et al. 2017). The study also showed that people aged 25 years and over with more than 30,000 mfs/mL died significantly earlier than amicrofilaremic subjects (Fig. 3). The second study, conducted in 2021 in 53 villages in the Republic of Congo, showed that the median survival time of amicrofilaremic subjects was 58.3 years compared to 39.2 years in *L. loa* microfilaremic subjects (Hemilembolo et al. 2023). Thus, loiasis clearly appears to be associated with excess mortality. In the Cameroon study, it was calculated that the population-attributable fraction of mortality associated with the presence of *L. loa* microfilaremia was 14.5%.

3.6 Burden of Loiasis

3.6.1 Burden of Loiasis as a Disease

Data from the 2017 study in Gabon allowed the authors to make the first estimate of the burden of loiasis in terms of DALYS (Veletzky et al. 2020). Considering only episodes of eyeworm, Calabar swellings, arthralgia, and severe headache, they calculated that loiasis was responsible for 412.9 DALYs per 100,000 inhabitants in rural areas of Gabon highly endemic for loiasis, and 82.2 DALYs per 100,000 inhabitants across the whole of Gabon; this latter value was close to that of the burden of genitourinary schistosomiasis in this country (103.3 DALYs per 100,000 inhabitants). Similar studies should be conducted in other foci of loiasis.

The burden related to the complications of loiasis is more difficult to assess due to the lack of data on the frequency and severity of these complications. Studies are underway to gather information and may help explain why loiasis is associated with excess mortality and allow more accurate estimates of the total burden of loiasis, including that due to frequent manifestations of the disease, complications, and excess mortality.

It is well known that the clinical and biological manifestations of loiasis are different in people exposed since childhood to the parasite and in travelers and expatriates exposed for the first time in adulthood (Klion et al. 1991; Churchill et al.

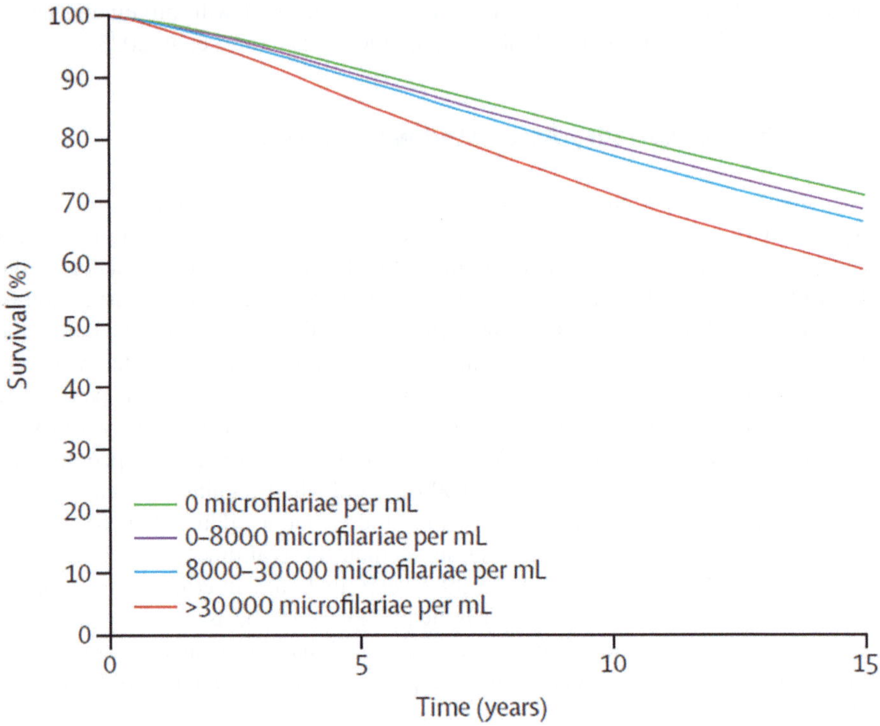

Fig. 3 Adjusted predicted survival curves according to the level of *Loa loa* MFD (four categories). Curves were fitted after the multivariable model for the older than 25 years of age group that showed the higher than 30,000 mfs/mL group was significantly associated with reduced survival. Figure extracted from Chesnais et al. (2017). *Lancet Infectious Diseases*, **17**(1): 108–116

1996; Antinori et al. 2012; Gantois et al. 2013). The latter, in particular, present more frequently with Calabar swellings than the former. To our knowledge, no studies have been conducted to determine whether there are differences in the clinical presentations in populations living in different endemic areas.

3.6.2 Economic Burden of Loiasis

The classical manifestations of loiasis can lead affected individuals to embark on a course of care aimed at relieving signs and symptoms. The cost of consultations and the purchase of medicines, modern or traditional, can be considerable and have a significant impact on household income (Ndzana 2011). Here again, specific studies must be carried out to assess this economic burden in regions where onchocerciasis is not coendemic, so as not to confuse the respective burdens of the two diseases, which may have similar manifestations (pruritus for example) and be confused in the local nosology.

People suffering from classic signs of loiasis can also suffer a significant economic burden due to difficulty in carrying out certain daily activities. For example, having a joint limited in movement for several days due to Calabar swelling can impair the ability to perform manual labor and thus have a definite impact on income.

3.6.3 Burden of Loiasis Related to the SAEs Occurring After Antifilarial Treatment

Ivermectin is widely distributed in central Africa to control onchocerciasis. This drug can cause moderate side effects in the setting of infection with *O. volvulus* and/or *L. loa*. These reactions (headache, pruritus, arthralgia, fever, etc.) generally last only a few days, during which activity may be reduced. In people with an *L. loa* MFD greater than 30,000 mfs/mL, ivermectin can lead to potentially fatal encephalopathy (Boussinesq et al. 1998, 2003). Between 1990 and 2017, more than 500 cases of post-ivermectin *L. loa*-related encephalopathy were reported to the Mectizan Donation Program (Kamgno et al. 2017). The number of deaths and the proportion of subjects with long-term sequelae after these SAEs are unknown. Fortunately, the number of cases in Cameroon and the DRC, the two countries that reported the highest numbers of SAEs, has dramatically decreased over the past 10 years (Makenga Bof et al. 2019).

Coendemicity of loiasis in onchocerciasis foci constitutes a significant burden for ivermectin distribution programs. In these coendemic areas, reinforced surveillance measures must be put in place for early identification and treatment of SAEs (Mectizan Expert Committee and Technical Consultative Committee 2004). Consequently, the cost per person treated is significantly higher than in areas where loiasis is absent (McFarland et al. 2005). Furthermore, population awareness of the occurrence of SAEs during distribution campaigns has led to therapeutic coverage that is often lower in coendemic areas (Haselow et al. 2003). Fear of side effects leads a relatively large proportion of people (sometimes $>15\%$) to be systematic non-compliers (Wanji et al. 2015b; Senyongo et al. 2016). The risk of SAEs also prevents extension of ivermectin distributions to areas where onchocerciasis is hypoendemic and loiasis is coendemic. These two phenomena (maintenance of a reservoir of parasites in onchocerciasis meso-hyperendemic areas and absence of treatment in hypoendemic zones) delay the elimination of onchocerciasis. Alternative treatment strategies can solve the problem of onchocerciasis-loiasis coendemicity, but they are more expensive to implement than community-directed treatment with ivermectin (Boussinesq et al. 2018).

For a long time, the risk of occurrence of SAEs delayed the launch of annual mass treatment programs against lymphatic filariasis (with the combination ivermectin +albendazole) in regions where this disease is coendemic with loiasis. Although such treatments reduce the transmission of *Wuchereria bancrofti,* they do not lead to clinical improvement in the infected subjects. Under these conditions, it was felt to be unethical to distribute ivermectin in areas endemic for loiasis given the risk of SAEs in the target populations. Fortunately, the demonstration that 6-monthly treatments with albendazole alone (a drug which has no microfilaricidal effect and which therefore does not induce SAE in people infected with *L. loa*) could eliminate lymphatic filariasis (Pion et al. 2022) led the WHO to recommend this alternative strategy and to launch mass distributions throughout central Africa (WHO 2017).

Here again, loiasis had an impact in terms of public health due to the delay in setting up the program.

4 Conclusions

There is growing evidence that loiasis is a disease that has a marked impact on the daily lives of sufferers and significant repercussions in terms of morbidity and mortality. Further studies are needed to clarify the frequency and severity of its complications and disease burden. This information should make it possible to include loiasis in the WHO list of neglected tropical diseases and to consider the implementation of specific control programs to combat it.

Acknowledgments I wish to thank Jean-Paul Akué, Jérémy Campillo, Cédric Chesnais, Amy Klion, Sébastien Pion, and Charles Whittaker for their useful comments on this text.

References

Abbate JL, Becquart P, Leroy E, Ezenwa VO, Roche B (2020) Exposure to Ebola virus and risk for infection with malaria parasites, rural Gabon. Emerg Infect Dis 26(2):229–237

Akue JP, Bivigou-Mboumba B, Padilla CM, Kazanji M, Nkoghe D (2020) Coinfection of human filarial *Loa loa, Mansonella perstans* in Human T Lymphotropic Virus Type 1 infected individuals in Gabon. Ann Infect Dis Clin Microbiol 1:100001

Akue JP, Devaney E (2002) Transmission intensity affects both antigen-specific and nonspecific T-cell proliferative responses in *Loa loa* infection. Infect Immun 70(3):1475–1480

Akue JP, Devaney E, Wahl G, Moukana H (2002) Expression of filarial-specific IgG subclasses under different transmission intensities in a region endemic for loiasis. Am J Trop Med Hyg 66(3):245–250

Akue JP, Hommel M, Devaney E (1997) High levels of parasite-specific IgG1 correlate with the amicrofilaremic state in *Loa loa* infection. J Infect Dis 175(1):158–163

Akue JP, Hommel M, Devaney E (1998) IgG subclass recognition of *Loa loa* antigens and their correlation with clinical status in individuals from Gabon. Parasite Immunol 20(8):387–393

Akue JP, Makouloutou P, Moutsimbi RM, Moukana H, Leroy E (2016) Influence of infective stage (L3) dose on the outcome of microfilaremia, peripheral white blood cells and humoral immune response in *Loa loa* experimentally infected *Mandrillus sphinx*. J Parasitol Vector Biol 8(4): 39–46

Akue JP, Nkoghe D, Padilla C, Moussavou G, Moukana H, Mbou RA, Ollomo B, Leroy EM (2011) Epidemiology of concomitant infection due to *Loa loa* and *Mansonella perstans* in Gabon. PLoS Negl Trop Dis 5(10):e1329

Antinori S, Schifanella L, Million M, Galimberti L, Ferraris L, Mandia L, Trabucchi G, Cacioppo V, Monaco G, Tosoni A, Brouqui P, Gismondo MR, Giuliani G, Corbellino M (2012) Imported Loa loa filariasis: three cases and a review of cases reported in non-endemic countries in the past 25 years. Int J Infect Dis 16(9):e649–e662

Austen EE (1909) Illustrations of African blood-sucking flies other than mosquitoes and tsetse-flies. British Museum (Natural History), London. 221 pp

Badia-Rius X, Betts H, Molyneux DH, Kelly-Hope LA (2019) Environmental factors associated with the distribution of *Loa loa* vectors *Chrysops* spp. in Central and West Africa: seeing the forest for the trees. Parasit Vectors 12:72

Bain O, Moisson P, Huerre M, Landsoud-Soukate J, Tutin C (1995) Filariae from a wild gorilla in Gabon with description of a new species of *Mansonella*. Parasite 2(3):315–322
Baird JK, Neafie RC, Connor DH (1988) Nodules in the conjunctiva, bung-eye, and bulge-eye in Africa caused by *Mansonella perstans*. Am J Trop Med Hyg 38(3):553–557
Baize S, Wahl G, Soboslay PT, Egwang TG, Georges AJ (1997) T helper responsiveness in human *Loa loa* infection; defective specific proliferation and cytokine production by CD4+ T cells from microfilaraemic subjects compared with amicrofilaraemics. Clin Exp Immunol 108(2):272–278
Bakajika DK, Nigo MM, Lotsima JP, Masikini GA, Fischer K, Lloyd MM, Weil GJ, Fischer PU (2014) Filarial antigenemia and *Loa loa* night blood microfilaremia in an area without bancroftian filariasis in the Democratic Republic of Congo. Am J Trop Med Hyg 91(6): 1142–1148
Blacklock DB (1930) *Report on a survey of human diseases in the Protectorate of Sierra Leone. Part I-Northern Province. Part II-Central and Southern Provinces*. Government printer. 44 pp
Bottieau E, Huits R, Van Den Broucke S, Maniewski U, Declercq S, Brosius I, Theunissen C, Feyens AM, Van Esbroeck M, van Griensven J, Clerinx J, Soentjens P (2022) Human filariasis in travelers and migrants: a retrospective 25-year analysis at the Institute of Tropical Medicine, Antwerp, Belgium. Clin Infect Dis 74(11):1972–1978
Boulesteix G, Carme B (1986) Encéphalite au cours du traitement de la filariose à *Loa loa* par la diéthylcarbamazine. A propos de 6 observations. Bulletin de la Société de Pathologie Exotique 79(5):649–654
Boussinesq M (2006) Loiasis. Ann Trop Med Parasitol 100(8):715–731
Boussinesq M, Fobi G, Kuesel AC (2018) Alternative treatment strategies to accelerate the elimination of onchocerciasis. Int Health 10(Suppl 1):i40–i48
Boussinesq M, Gardon J (1997) Prevalences of *Loa loa* microfilaraemia throughout the area endemic for the infection. Ann Trop Med Parasitol 91(6):573–589
Boussinesq M, Gardon J, Gardon-Wendel N, Chippaux JP (2003) Clinical picture, epidemiology and outcome of *Loa*-associated serious adverse events related to mass ivermectin treatment of onchocerciasis in Cameroon. Filaria J 2(Suppl 1):S4
Boussinesq M, Gardon J, Gardon-Wendel N, Kamgno J, Ngoumou P, Chippaux JP (1998) Three probable cases of *Loa loa* encephalopathy following ivermectin treatment for onchocerciasis. Am J Trop Med Hyg 58(4):461–469
Boussinesq M, Gardon J, Kamgno J, Pion SDS, Gardon-Wendel N, Chippaux JP (2001) Relationships between the prevalence and intensity of *Loa loa* infection in the Central province of Cameroon. Ann Trop Med Parasitol 95(5):495–507
Bouyou-Akotet MK, Boussougou MM, Ovono-Abessolo F, Owono-Medang M, Kombila M (2014) Influence of *Mansonella perstans* microfilaraemia on total IgE levels in Gabonese patients co-infected with *Loa loa*. Acta Trop 131:11–15
Buckley JJC (1946) A helminthological survey in Northern Rhodesia. J Helminthol 21(4):111–174
Buell KG, Whittaker C, Chesnais CB, Jewell PD, Pion SD, Walker M, Basanez MG, Boussinesq M (2019) Atypical clinical manifestations of loiasis and their relevance for endemic populations. Open Forum Infect Dis 6(11):ofz417
Campillo JT, Hemilembolo MC, Louya F, Bikita P, Pion SD, Boussinesq M, Missamou F, Chesnais CB (2023a) Temporal variability of *Loa loa* microfilaraemia. Parasit Vectors 16:23
Campillo JT, Hemilembolo MC, Pion SDS, Lebredonchel E, Dupasquier V, Boullé C, Rancé LG, Boussinesq M, Missamou F, Chesnais CB (2023b) Association between blood *Loa loa* microfilarial density and proteinuria levels: a population based cross-sectional study in a rural area of the Republic of Congo. Lancet Microbe. in press
Carme B, Mamboueni JP, Copin N, Noireau F (1989) Clinical and biological study of *Loa loa* filariasis in Congolese. Am J Trop Med Hyg 41(3):331–337
Casaca VMR (1966) Contribuição para o estudo das filaríases existentes em Angola. 1. *Onchocerca volvulus*, 2. *Loa loa*, 3. *Dipetalonema perstans*. Anais do Instituto de Medicina Tropical 23(1-2):133–149

Caubère P, Noireau F (1991) Effect of attraction factors on the sampling of *Chrysops silacea* and *C. dimidiata* (Diptera: Tabanidae), vectors of *Loa loa* (Filaroidea: Onchocercidae) filariasis. J Med Entomol 28(2):263–265

Checkouri T, Missamou F, Pion SDS, Bikita P, Hemilembolo MC, Boussinesq M, Chesnais CB, Campillo JT (2023) Association between altered cognition and *Loa loa* microfilaremia: First evidence from a cross-sectional study in a rural area of the Republic of Congo. PLoS Negl Trop Dis 17(6):e0011430

Chesnais CB, Takougang I, Paguélé M, Pion SD, Boussinesq M (2017) Excess mortality associated with loiasis: a retrospective population-based cohort study. Lancet Infect Dis 17(1):108–116

Chhabra RC, Bhat S, Shukla SM (1989) Occular loiasis in a Zambian woman. East Afr Med J 66(7): 491–494

Chippaux JP, Bouchité B, Demanou M, Morlais I, Le Goff G (2000) Density and dispersal of the loaiasis vector *Chrysops dimidiata* in southern Cameroon. Med Vet Entomol 14(3):339–344

Chippaux JP, Boussinesq M, Gardon J, Gardon-Wendel N, Ernould JC (1996) Severe adverse reaction risks during mass treatment with ivermectin in loiasis-endemic areas. Parasitol Today 12(11):448–450

Churchill DR, Morris C, Fakoya A, Wright SG, Davidson RN (1996) Clinical and laboratory features of patients with loiasis (*Loa loa* filariasis) in the UK. J Infect 33(2):103–109

Corson JF, Ingram A (1923) Reports on the Medical Research Institute from January 1st, 1922, to March 31st, 1923, 13 pp

Cot M, Le Hesran JY, Miailhes P, Cot S, Hougard JM, Froment A (1995) Indicateurs de santé dans la population d'un complexe agro-industriel du Sud Cameroun. Cahiers Santé 5(3):167–180

Crewe W, Williams P (1961) The bionomics of the tabanid fauna of streams in the rain-forest of the Southern Cameroons: I. Oviposition. Ann Trop Med Parasitol 55(3):363–378

Crosskey RW, Cogan BH, Freeman P, Pont AC, Smith KGV, Oldroyd H (1980) Catalogue of the Diptera of the Afrotropical region. British Museum (Natural History), London. 1437 pp

Davey JT, O'Rourke FJ (1951) Observations on *Chrysops silacea* and *C. dimidiata* at Benin, Southern Nigeria. Part I. Ann Trop Med Parasitol 45(1):30–37

Delaporte E, Peeters M, Durand JP, Dupont A, Schrijvers D, Bedjabaga L, Honoré C, Ossari S, Trébucq A, Josse R, Merlin M (1989) Seroepidemiological survey of HTLV-I infection among randomized populations of western central African countries. J Acquir Immune Defic Syndr 2(4):410–414

Demanou M, Pion SDS, Boussinesq M (2001) Etude entomologique sur la transmission de la loase dans le département de la Lékié (Cameroun). Bulletin de la Société de Pathologie exotique 94(4):347–352

Dennis VA, Lowrie RC Jr, Osae-Addo G, Blanchard JL (1993) Microfilarial densities, hematologic changes, and serum antibody levels in *Loa loa*-infected rhesus monkeys (*Macaca mulatta*). Am J Trop Med Hyg 49(6):763–771

Develoux M, Hennequin C, Le Loup G, Paris L, Magne D, Belkadi G, Pialoux G (2017) Imported filariasis in Europe: a series of 31 cases from Metropolitan France. Eur J Intern Med 37:e37–e39

Diggle PJ, Thomson MC, Christensen OF, Rowlingson B, Obsomer V, Gardon J, Wanji S, Takougang I, Enyong P, Kamgno J, Remme JH, Boussinesq M, Molyneux DH (2007) Spatial modelling and the prediction of *Loa loa* risk: decision making under uncertainty. Ann Trop Med Parasitol 101(6):499–509

Drame PM, Montavon C, Pion SD, Kubofcik J, Fay MP, Nutman TB (2016) Molecular epidemiology of blood-borne human parasites in a *Loa loa*-, *Mansonella perstans*-, and *Plasmodium falciparum*-endemic region of Cameroon. Am J Trop Med Hyg 94(6):1301–1308

Duke BOL (1955a) Studies on the biting habits of *Chrysops*: II. The effect of wood fires on the biting density of *Chrysops silacea* in the rain-forest at Kumba, British Cameroons. Ann Trop Med Parasitol 49(3):260–272

Duke BOL (1955b) Symposium on loiasis. IV. The development of *Loa* in flies of the genus *Chrysops* and the probable significance of the different species in the transmission of loiasis. Trans R Soc Trop Med Hyg 49(2):115–121

Duke BOL (1957) Experimental transmission of *Loa loa* from man to monkey. Nature 179(4574): 1357–1358

Duke BOL (1959) Studies on the biting habits of *Chrysops*: VI. A comparison of the biting habits, monthly biting densities and infection rates of *C. silacea* and *C. dimidiata* (Bombe form) in the rain-forest at Kumba, Southern Cameroons, UUKA. Ann Trop Med Parasitol 53(2):203–214

Duke BOL (1964) Studies on loiasis in monkeys: IV. Experimental hybridization of the human and simian strains of *Loa*. Ann Trop Med Parasitol 58(4):390–408

Duke BOL, Wijers DJB (1958) Studies on loiasis in monkeys: I. The relationship between human and simian *Loa* in the rain-forest zone of the British Cameroons. Ann Trop Med Parasitol 52(2): 158–175

Eberhard ML, Orihel TC (1981) Development and larval morphology of *Loa loa* in experimental primate hosts. J Parasitol 67(4):556–564

Emukah E, Rakers LJ, Kahansim B, Miri ES, Nwoke BE, Griswold E, Saka Y, Anagbogu I, Davies E, Ityonzughul C, D'Ambrosio M, Bakalar M, Fletcher DA, Nutman T, Kamgno J, Richards FO Jr (2018) In southern Nigeria *Loa loa* blood microfilaria density is very low even in areas with high prevalence of loiasis: results of a survey using the new LoaScope technology. Am J Trop Med Hyg 99(1):116–123

Eyebe S, Sabbagh A, Pion SD, Nana-Djeunga HC, Kamgno J, Boussinesq M, Chesnais CB (2018) Familial aggregation and heritability of *Loa loa* microfilaremia. Clin Infect Dis 66(5):751–757

Fain A (1978) Les problèmes actuels de la loase. Bull World Health Organ 56(2):155–167

Fain A, Elsen P, Wéry M, Maertens K (1974) Les filarioses humaines au Mayumbe et dans les régions limitrophes (République du Zaïre). Evaluation de la densité microfilarienne. Ann Soc Belg Med Trop 54(1):5–34

Fülleborn F (1908) Untersuchungen an menschlichen Filarien und deren Uebertragung auf Stechmücken. Archiv für Schiffs- und Tropenhygiene 12(9):356–388

Gantois N, Rapp C, Gautret P, Ficko C, Savini H, Larreché S, Saidi R, Crevon L, Simon F (2013) Imported loiasis in France: a retrospective analysis of 47 cases. Travel Med Infect Dis 11(6): 366–373

Garcia A, Abel L, Cot M, Ranque S, Richard P, Boussinesq M, Chippaux JP (1995) Longitudinal survey of *Loa loa* filariasis in southern Cameroon: long-term stability and factors influencing individual microfilarial status. Am J Trop Med Hyg 52(4):370–375

Garcia A, Abel L, Cot M, Richard P, Ranque S, Feingold J, Demenais F, Boussinesq M, Chippaux JP (1999) Genetic epidemiology of host predisposition microfilaraemia in human loiasis. Trop Med Int Health 4(8):565–574

Gardon J, Gardon-Wendel N, Demanga-Ngangue, Kamgno J, Chippaux JP, Boussinesq M (1997) Serious reactions after mass treatment of onchocerciasis with ivermectin in an area endemic for *Loa loa* infection. Lancet 350(9070):18–22

Gordon RM, Kershaw WE, Crewe W, Oldroyd H (1950) The problem of loiasis in West Africa with special reference to recent investigations at Kumba in the British Cameroons and at Sapele in Southern Nigeria. Trans R Soc Trop Med Hyg 44(1):11–47

Gouteux JP, Noireau F, Staak C (1989) The host preferences of *Chrysops silacea* and *C. dimidiata* (Diptera: Tabanidae) in an endemic area of *Loa loa* in the Congo. Ann Trop Med Parasitol 83(2):167–172

Gryseels B, Polderman AM, Manshande JP, Gigase PL (1985) Human filariasis in two forest villages in Maniema (Kivu, Zaire). Ann Soc Belg Med Trop 65(2):163–171

Haselow NJ, Akame J, Evini C, Akongo S (2003) Programmatic and communication issues in relation to serious adverse events following ivermectin treatment in areas co-endemic for onchocerciasis and loiasis. Filaria J 2(Suppl 1):S10

Hawking F (1977) The distribution of human filariasis throughout the world. Part III. Africa. Trop Dis Bull 74(8):649–679

Hemilembolo MC, Niama AC, Campillo JT, Pion SD, Missamou F, Whittaker C, Kankou JM, Ndziessi G, Bileckot RR, Boussinesq M, Chesnais CB (2023) Excess mortality associated with

loiasis: confirmation by a new retrospective cohort study conducted in the Republic of Congo. *Open Forum*. Infect Dis 10(3):ofad103
Hughes MH, Sarkies JWR (1951) The length of exposure to infestation and the danger of contracting onchocerciasis. Ann Trop Med Parasitol 45(1):73–77
Janssens PG, Van Bogaert L, Tverdy G, Wanson M (1958) Réflexions sur le sort des microfilaires de *Loa loa* dans l'organisme humain parasité. Manifestations viscérales provoquées par leur infiltration dans les tissus. Bulletin de la Société de Pathologie Exotique 51(4):632–645
Jazuli F, Kelton TJ, Keystone JS (2016) A diagnostic challenge: eosinophilia of unknown etiology. Travel Med Infect Dis 5(14):537–538
Johnson O, Giorgi E, Fronterrè C, Amoah B, Atsame J, Ella SN, Biamonte M, Ogoussan K, Hundley L, Gass K, Diggle PJ (2022) Geostatistical modelling enables efficient safety assessment for mass drug administration with ivermectin in *Loa loa* endemic areas through a combined antibody and LoaScope testing strategy for elimination of onchocerciasis. PLoS Negl Trop Dis 16(2):e0010189
Johnstone RDC (1947) Loiasis. Lancet 1(6442):250–253
Kamgno J, Pion SD, Chesnais CB, Bakalar MH, D'Ambrosio MV, Mackenzie CD, Nana-Djeunga HC, Gounoue-Kamkumo R, Njitchouang GR, Nwane P, Tchatchueng-Mbouga JB, Wanji S, Stolk WA, Fletcher DA, Klion AD, Nutman TB, Boussinesq M (2017) A test-and-not-treat strategy for onchocerciasis in *Loa loa*–endemic areas. N Engl J Med 377(21):2044–2052
Kelly-Hope LA, Bockarie MJ, Molyneux DH (2012) *Loa loa* ecology in central Africa: role of the Congo River system. PLoS Negl Trop Dis 6(6):e1605
Kelly-Hope LA, Cano J, Stanton MC, Bockarie MJ, Molyneux DH (2014) Innovative tools for assessing risks for severe adverse events in areas of overlapping *Loa loa* and other filarial distributions: the application of micro-stratification mapping. Parasit Vectors 7:307
Kelly-Hope L, Paulo R, Thomas B, Brito M, Unnasch TR, Molyneux D (2017) *Loa loa* vectors *Chrysops* spp.: perspectives on research, distribution, bionomics, and implications for elimination of lymphatic filariasis and onchocerciasis. Parasit Vectors 10:172
Kershaw WE, Crewe W, Beesley WN (1954) Studies on the intake of microfilariae by their insect vectors, their survival, and their effect on the survival of their vectors. II. The intake of the microfilariae of *Loa loa* and *Acanthocheilonema perstans* by *Chrysops* spp. Ann Trop Med Parasitol 48(1):102–109
Kershaw WE, Duke BOL (1954) Studies on the intake of microfilariae by their insect vectors, their survival, and their effect on the survival of their vectors. V. The survival of *Loa loa* in *Chrysops silacea* under laboratory conditions. Ann Trop Med Parasitol 48(3):340–344
Kershaw WE, Keay RWJ, Nicholas WL, Zahra A (1953) Studies on the epidemiology of filariasis in West Africa, with special reference to the British Cameroons and the Niger delta. IV. The incidence of *Loa loa* and *Acanthocheilonema perstans* in the rain-forest, the forest fringe and the mountain grasslands of the British Cameroons, with observations on the species of *Chrysops* and *Culicoides* found. Ann Trop Med Parasitol 47(4):406–425
Klion AD, Massougbodji A, Sadeler BC, Ottesen EA, Nutman TB (1991) Loiasis in endemic and nonendemic populations: immunologically mediated differences in clinical presentation. J Infect Dis 163(6):1318–1325
Kumari V, Ahmad S, Singh A, Banerjee T (2019) Presence of adult *Loa loa* in the anterior chamber of the eye along with microfilaremia from nonendemic region: a rare presentation from India. Tzu-Chi Med J 31(4):283–285
Léger A (1912) La filariose humaine dans le Haut-Sénégal et Niger. Index endémique de la région de Bamako. Bulletin de la Société de Pathologie Exotique 5(8):618–622
Leiper RT (1913) Reports from the London School of Tropical Medicine: report of the helminthologist for the half-year ending April 30th, 1913. Report of the Advisory Committee for the Tropical Diseases Research Fund for the year 1913, pp 86–87
Lipner EM, Law MA, Barnett E, Keystone JS, von Sonnenburg F, Loutan L, Prevots DR, Klion AD, Nutman TB, GeoSentinel Surveillance Network (2007) Filariasis in travelers presenting to the GeoSentinel Surveillance Network. PLoS Negl Trop Dis 1(3):e88

Loiseau C, Gutiérrez-López R, Mathieu B, Makanga BK, Paupy C, Rahola N, Cornel AJ (2022) Diversity and distribution of the arthropod vectors of the Gulf of Guinea oceanic islands. In: Ceríaco et al (eds) Biodiversity of the Gulf of Guinea oceanic islands. Springer, Cham, pp 383–405

Makenga Bof JC, Muteba D, Mansiangi P, Ilunga-Ilunga F, Coppieters Y (2019) Analysis of severe adverse effects following community-based ivermectin treatment in the Democratic Republic of Congo. BMC Pharmacol Toxicol 20:49

Mandal D, Roy D, Bera DK, Manna B (2013) Occurrence of gravid *Loa loa* in subconjunctival space of man: a case report from West Bengal, India. J Parasit Dis 37(1):52–55

Maplestone PA (1938) A new filarial worm from a human being. Ind Medical Gazette 73(1):8–10

Marriott WR (1986) Loiasis in a young child in Oregon. Int J Dermatol 25(4):252–254

McFarland D, Menzies N, Njoumemi Z, Onwujekwe O (2005) Study of cost per treatment with ivermectin using the CDTI strategy. African Programme for Onchocerciasis Control (APOC)

Mectizan® Expert Committee and Technical Consultative Committee (2004) Recommendations for the treatment of onchocerciasis with Mectizan® in areas co-endemic for onchocerciasis and loiasis. Atlanta, Mectizan® Donation Program, 2004

Mengome LE, Mewono L, Aboughe-Angone S (2018) Are medicinal plants the future of *Loa loa* treatment? Pharmacogn Rev 12(23):133–137

Mischlinger J, Veletzky L, Tazemda-Kuitsouc GB, Pitzinger P, Matsegui PB, Gmeiner M, Lagler H, Gebru T, Held J, Mordmüller B, Ramharter M (2018) Behavioural and clinical predictors for loiasis. J Glob Health 8(1):010413

Mommers EC, Dekker HS, Richard P, Garcia A, Chippaux JP (1995) Prevalence of *L. loa* and *M. perstans* filariasis in southern Cameroon. Trop Geogr Med 47(1):2–5

Moraga P, Cano J, Baggaley RF, Gyapong JO, Njenga SM, Nikolay B, Davies E, Rebollo MP, Pullan RL, Bockarie MJ, Hollingsworth TD, Gambhir M, Brooker SJ (2015) Modelling the distribution and transmission intensity of lymphatic filariasis in sub-Saharan Africa prior to scaling up interventions: integrated use of geostatistical and mathematical modelling. Parasit Vectors 8:560

Ndzana IB (2011) La loase comme maladie clinique et culturelle au Cameroun. In: Nkoum BA (ed) Santé plurielle en Afrique : Perspective pluridisciplinaire. L'Harmattan, Paris, pp 205–223

Nelson GS (1965) Filarial infections as zoonoses. J Helminthol 39(2–3):229–250

Nicolini AL, Tamarozzi F, Pomari E, Mistretta M, Camera M, Sepulcri C, Bassetti M, Gobbi FG (2022) Loiasis from where you don't expect it: an illustrative case of misled diagnosis. J Travel Med 29(7):taac060

Njambe Priso GD, Lissom A, Ngu LN, Nji NN, Tchadji JC, Tchouangueu TF, Ambada GE, Sake Ngane CS, Dafeu BL, Djukouo L, Nyebe I, Magagoum S, Ngoh AA, Ouambo Fotso H, Garcia R, Guttiérrez A, Okoli AS, Esimone CO, Njiokou F, Park CG, Bopda Waffo A, Nchinda GW (2018) Filaria specific antibody response profiling in plasma from anti-retroviral naïve *Loa loa* microfilaraemic HIV-1 infected people. BMC Infect Dis 18:160

Nnochiri E (1972) The causal agent of the ocular syndrome of the "Kampala eye worm". East Afr Med J 49:198

Noireau F, Apembet JD, Nzoulani A, Carme B (1990a) Clinical manifestations of loiasis in an endemic area in the Congo. Trop Med Parasitol 41(1):37–39

Noireau F, Carme B, Apembet JD, Gouteux JP (1989) *Loa loa* and *Mansonella perstans* filariasis in the Chaillu Mountains, Congo: parasitological prevalence. Trans R Soc Trop Med Hyg 83(4): 529–534

Noireau F, Gouteux JP (1989) Current considerations on a *Loa loa* simian reservoir in the Congo. Acta Trop 46(1):69–70

Noireau F, Nzoulani A, Sinda D, Caubere P (1991) *Chrysops silacea* and *C. dimidiata* seasonality and loiasis prevalence in the Chaillu Mountains, Congo. Med Vet Entomol 5(4):413–419

Noireau F, Nzoulani A, Sinda D, Itoua A (1990b) Transmission indices of *Loa loa* in the Chaillu Mountains, Congo. Am J Trop Med Hyg 43(3):282–288

Noireau F, Pichon G (1992) Population dynamics of *Loa loa* and *Mansonella perstans* infections in individuals living in an endemic area of the Congo. Am J Trop Med Hyg 46(6):672–676
Ogunba EO (1972) The development of *Loa loa* (Guyot) in *Mansonia africana* (Theobald). J Med Entomol 9(2):159–161
Ogunba EO (1977) *Loa-loa* microfilaremia in the Nigerian population. Trop Geogr Med 29(1): 51–55
Oldroyd H (1957) The horse-flies (Diptera: Tabanidae) of the Ethiopian Region. III. Subfamilies Chrysopinae, Scepsidinae and Pangoniinae and a revised classification. British Museum (Natural History), London. xii + 489 pp
Orihel TC, Eberhard ML (1985) *Loa loa*: development and course of patency in experimentally-infected primates. Trop Med Parasitol 36(4):215–224
Orihel TC, Eberhard ML (1998) Zoonotic filariasis. Clin Microbiol Rev 11(2):366–381
Orihel TC, Lowrie RC Jr (1975) *Loa loa*: development to the infective stage in an American deerfly, *Chrysops atlanticus*. Am J Trop Med Hyg 24(4):610–615
Orihel TC, Moore PJ (1975) *Loa loa*: experimental infection in two species of African primates. Am J Trop Med Hyg 24(4):606–609
Pedram B, Pasquetto V, Drame PM, Ji Y, Gonzalez-Moa MJ, Baldwin RK, Nutman TB, Biamonte MA (2017) A novel rapid test for detecting antibody responses to *Loa loa* infections. PLoS Negl Trop Dis 11(7):e0005741
Pinder M (1988) *Loa loa*—a neglected filaria. Parasitol Today 4(10):279–284
Pinder M, Everaere S, Roelants GE (1994) *Loa loa*: immunological responses during experimental infections in mandrills (*Mandrillus sphinx*). Exp Parasitol 79(2):126–136
Pinder M, Leclerc A, Flockhart HA, Egwang TG (1990) *Macaca fascicularis*, a nonpermissive host for the human filarial parasite *Loa loa*. J Parasitol 76(3):373–376
Pion SD, Chesnais CB, Weil GJ, Louya F, Boussinesq M, Missamou F (2022) Impact of semi-annual albendazole on lymphatic filariasis and soil-transmitted helminth infection: parasitological assessment after 14 rounds of community treatment. Am J Trop Med Hyg 106(2):729–731
Pion SDS, Clarke P, Filipe JAN, Kamgno J, Gardon J, Basanez MG, Boussinesq M (2006a) Co-infection with *Onchocerca volvulus* and *Loa loa* microfilariae in central Cameroon: are these two species interacting? Parasitology 132(6):843–854
Pion SDS, Demanou M, Oudin B, Boussinesq M (2005) Loiasis: the individual factors associated with the presence of microfilaraemia. Ann Trop Med Parasitol 99(5):491–500
Pion SDS, Filipe JAN, Kamgno J, Gardon J, Basánez MG, Boussinesq M (2006b) Microfilarial distribution of *Loa loa* in the human host: population dynamics and epidemiological implications. Parasitology 133(1):101–109
Pion SDS, Gardon J, Kamgno J, Gardon-Wendel N, Chippaux JP, Boussinesq M (2004) Structure of the microfilarial reservoir of *Loa loa* in the human host and its implications for monitoring the programmes of Community-Directed Treatment with Ivermectin carried out in Africa. Parasitology 129(5):613–626
Pion SD, Njiokou F, Boussinesq M, Simo G, Truc P (2017) Confirmation of no influence of *Loa loa* and *Mansonella perstans* on the Card Agglutination Test for Trypanosomosis used for serological screening of Human African Trypanosomosis. J Trop Dis 5(4):1000242
Poindexter HA (1953) An analytical study of 45,000 consecutive admissions to a clinic in Monrovia, Liberia, West Africa. J Natl Med Assoc 45(5):345–349
Poltera AA (1973) The histopathology of ocular loiasis in Uganda. Trans R Soc Trop Med Hyg 67(6):819–829
Pongui Ngondza B, Koumba Lengongo JV, Mickala P, M'bondoukwé NP, Ndong Ngomo JM, Moutombi Ditombi BC, Mawili-Mboumba DP, Bouyou-Akotet MK (2022) Prevalence and risk factors for blood filariasis among HIV-infected adults in Gabon, Central Africa: a pilot study. Trans R Soc Trop Med Hyg 116(11):1015–1021
Price DL (1961) The occurrence of *Loa loa* in Uganda. Trans R Soc Trop Med Hyg 55(2):199
Priest DH, Nutman TB (2017) Loiasis in US traveler returning from Bioko Island, Equatorial Guinea, 2016. Emerg Infect Dis 23(1):160–162

Puente S, Ramírez-Olivencia G, Lago M, Subirats M, Bru F, Pérez-Blazquez E, Arsuaga M, de Guevara CL, de la Calle-Prieto F, Vicente B, Alonso-Sardón M, Belhassen-Garcia M, Muro A (2020) Loiasis in sub-Saharan migrants living in Spain with emphasis of cases from Equatorial Guinea. Infect Dis Poverty 9:16

Rhodain F, Rodhain-Rebourg F (1973) A propos de la distribution géographique de la loase. Med Mal Infect 3(11):429–436

Richard A, Lallemant M, Trape JF, Carnevale P, Mouchet J (1988) Le paludisme dans la région forestière de Mayombe, République Populaire du Congo. III. Place du paludisme dans la morbidité générale. Annales de la Société Belge de Médecine Tropicale 68(4):317–329

Richardson ET, Luo R, Fink DL, Nutman TB, Geisse JK, Barry M (2012) Transient facial swellings in a patient with a remote African travel history. J Travel Med 19(3):183–185

Ripert C, Ambroise-Thomas P, Riedel D, Rousselle-Sauer C, Zimflou A, Ibrahima H (1977) Epidémiologie des filarioses à *L. loa* et *D. perstans* dans sept villages de la province du Centre-sud du Cameroun. Bulletin de la Société de Pathologie exotique 70(5):504–515

Rodhain F (1980) Hypothèses concernant l'écologie dynamique des infections à *Loa*. Bulletin de la Société de Pathologie exotique 73(2):182–191

Rodhain J (1946) Corollaire à l'étude de E. Peel et M. Chardome sur les filaridés des chimpanzés au Congo belge. Annales de la Société Belge de Médecine Tropicale 26(3):157–160

Rodhain J, Van den Berghe L (1939) *Paraloa anthropopitheci*, genre et espèce nouveaux de Filaroidea chez le chimpanzé au Congo Belge. Ann Soc Belg Med Trop 19(3):445–452

Sandground JH (1936) On the occurrence of a species of *Loa* in monkeys in the Belgian Congo. Ann Soc Belg Med Trop 16(2):273–278

Sandri TL, Kreidenweiss A, Cavallo S, Weber D, Juhas S, Rodi M, Woldearegai TG, Gmeiner M, Veletzky L, Ramharter M, Tazemda-Kuitsouc GB, Matsiegui PB, Mordmüller B, Held J (2021) Molecular epidemiology of *Mansonella* species in Gabon. J Infect Dis 223(2):287–296

Schlüter DK, Ndeffo-Mbah ML, Takougang I, Ukety T, Wanji S, Galvani AP, Diggle PJ (2016) Using community-level prevalence of *Loa loa* infection to predict the proportion of highly-infected individuals: statistical modelling to support lymphatic filariasis and onchocerciasis elimination programs. PLoS Negl Trop Dis 10(12):e0005157

Senyonjo L, Oye J, Bakajika D, Biholong B, Tekle A, Boakye D, Schmidt E, Elhassan E (2016) Factors associated with ivermectin non-compliance and its potential role in sustaining *Onchocerca volvulus* transmission in the west region of Cameroon. PLoS Negl Trop Dis 10(8):e0004905

Sharp NA (1929) *Loa loa* infections. A case with rapid onset of symptoms. Lancet 214(5537): 765–766

Strong RP, Shattuck GC, Theiler M, Whitman L, Bequaert JC, Allen GM, Linder DH, Coolidge HJ (1930) The African Republic of Liberia and the Belgian Congo. Based on the observations made and material collected during the Harvard African Expedition 1926-1927, vol I. Harvard University Press, Cambridge, 568 pp

Takougang I, Meli J, Lamlenn S, Tatah PN, Ntep M (2007) Loiasis—a neglected and under-estimated affliction: endemicity, morbidity and perceptions in eastern Cameroon. Ann Trop Med Parasitol 101(2):151–160

Takougang I, Meremikwu M, Wandji S, Yenshu EV, Aripko B, Lamlenn SB, Eka BL, Enyong P, Meli J, Kale O, Remme JH (2002) Rapid assessment method for prevalence and intensity of *Loa loa* infection. Bull World Health Organ 80(11):852–858

Taufflieb R, Finelle P (1956) Étude écologique et biologique des tabanidés d'Afrique Équatoriale Française. Bulletin Institut d'études centrafricaines 12:209–251

Tenaguem J, Wawo-Yonta E, Ngoumou Fotsing R, Kamtchum Tatuene J, Nana Djeunga H, Bopda J, Hopkins A, Nutman TB, Kuaban C, Boussinesq M, Klion A, Kamgno J (2010) Cardiac lesions in an area hyperendemic for loiasis in Cameroun. Am J Trop Med Hyg 83 (Suppl 5):396–397

Thomas J, Chastel C, Forcain L (1970) Latence clinique et parasitaire dans les filarioses à *Loa loa* et à *Onchocerca volvulus*. Bulletin de la Société de Pathologie Exotique 63(1):90–94

Thomson MC, Obsomer V, Kamgno J, Gardon J, Wanji S, Takougang I, Enyong P, Remme JH, Molyneux DH, Boussinesq M (2004) Mapping the distribution of *Loa loa* in Cameroon in support of the African Programme for Onchocerciasis Control. Filaria J 3:7

Thorpe VG (1892) The African *filaria sanguinis hominis*. Br Med J 1(1623):305–306

Touré FS, Deloron P, Egwang TG, Wahl G (1999) Relation entre intensité de la transmission de la filaire *Loa loa* et prévalence des infections. Med Trop 59(3):249–252

Touré FS, Mavoungou E, Kassambara L, Williams T, Wahl G, Millet P, Egwang TG (1998) Human occult loiasis: field evaluation of a nested polymerase chain reaction assay for the detection of occult infection. Trop Med Int Health 3(6):505–511

Treadgold CH (1920) On a filaria, *Loa papionis* n. sp., parasitic in *Papio cynocephalus*. Parasitology 12(2):113–135

Uttah EC (2013) The epidemiological implications of deer fly nuisance biting and transmission of loiasis in an endemic area in Southeastern Nigeria. Int Scholarly Res Notices 2013:970386

Van den Berghe L, Chardome M, Peel E (1957) Une microfilaire de type *Loa* chez *Gorilla gorilla* au Congo Belge. Folia scientifica Africae centralis 3(4):88

Van den Berghe L, Chardome M, Peel E (1964) The filarial parasites of the eastern gorilla in the Congo. J Helminthol 38(3-4):349–368

Veletzky L, Hergeth J, Stelzl DR, Mischlinger J, Manego RZ, Mombo-Ngoma G, McCall MBB, Adegnika AA, Agnandji ST, Metzger WG, Matsiegui PB, Lagler H, Mordmüller B, Budke C, Ramharter M (2020) Burden of disease in Gabon caused by loiasis: a cross-sectional survey. Lancet Infect Dis 20(11):1339–1346

Veletzky L, Mombo-Ngoma G, Ramharter M (2021) The supposedly innocuous eye worm: or, what could make you put chili in your eye? Am J Trop Med Hyg 105(6):1448–1449

Vinkeles Melchers NVS, Coffeng LE, Boussinesq M, Pedrique B, Pion SDS, Tekle AH, Zouré HGM, Wanji S, Remme JH, Stolk WA (2020) Projected number of people with onchocerciasis–loiasis coinfection in Africa, 1995 to 2025. Clin Infect Dis 70(11):2281–2289

Vogel H (1927) Beiträge zur Anatomie der Gattungen *Dirofilaria* und *Loa*. Zentralblatt fur Bakteriologie, Parasitenkunde, Infektionskrankheiten und Hygiene, Abt. I (Originale) 102(1–3):81–89

Volpicelli L, De Angelis M, Morano A, Biliotti E, Franchi C, Gabrielli S, Mattiucci S, Di Bonaventura C, Taliani G (2020) Encephalopathy in a patient with loiasis treated with albendazole: a case report. Parasitol Int 75:102006

Wanji S (2001) Rapid assessment procedures for loiasis: report of a multi-centre study. UNDP/World Bank/WHO Special Programme for Research and Training in Tropical Diseases (TDR), Geneva, TDR/IDE/RP/RAPL/01.1

Wanji S, Akotshi DO, Mutro MN, Tepage F, Ukety TO, Diggle PJ, Remme JH (2012) Validation of the rapid assessment procedure for loiasis (RAPLOA) in the Democratic Republic of Congo. Parasit Vectors 5(1):25

Wanji S, Eyong EE, Tendongfor N, Ngwa C, Esuka E, Kengne-Ouafo A, Datchoua-Poutcheu F, Enyong P, Hopkins A, Mackenzie CD (2015a) Parasitological, hematological and biochemical characteristics of a model of hyper-microfilariaemic loiasis (*Loa loa*) in the baboon (*Papio anubis*). PLoS Negl Trop Dis 9(11):e0004202

Wanji S, Kengne-Ouafo JA, Esum ME, Chounna PW, Adzemye BF, Eyong JE, Jato I, Datchoua-Poutcheu FR, Abong RA, Enyong P, Taylor DW (2015b) Relationship between oral declaration on adherence to ivermectin treatment and parasitological indicators of onchocerciasis in an area of persistent transmission despite a decade of mass drug administration in Cameroon. Parasit Vectors 8:667

Wanji S, Tendongfor N, Esum M, Atanga SN, Enyong P (2003) Heterogeneity in the prevalence and intensity of loiasis in five contrasting bioecological zones in Cameroon. Trans R Soc Trop Med Hyg 97(2):182–187

Wanji S, Tendongfor N, Esum ME, Enyong P (2002) *Chrysops silacea* biting densities and transmission potential in an endemic area of human loiasis in south-west Cameroon. Tropical Med Int Health 7(4):371–377

White GB (1977) Man-biting species of *Chrysops* Meigen, *Culicoides* Latreille and *Simulium* Latreille in Ethiopia, with discussion of their vector potentialities. Trans R Soc Trop Med Hyg 71(2):161–175

Whittaker C, Walker M, Pion SD, Chesnais CB, Boussinesq M, Basáñez MG (2018) The population biology and transmission dynamics of *Loa loa*. Trends Parasitol 34(4):335–350

Willcox RR (1945) Genital oedema in relation to filariasis in the Gold Coast. Br J Venereal Dis 21(4):178–179

Woodman HM, Bokhari A (1941) Studies on *Loa loa* and the first report of *Wuchereria bancrofti* in the Sudan. Trans R Soc Trop Med Hyg 35(2):77–92

World Health Organization (2017) Guideline: Alternative mass drug administration regimens to eliminate lymphatic filariasis. WHO/HTM/NTD/PCT/2017.07, Geneva

Wright JC (1961) A survey of medical conditions in Ghana in 1957. J Natl Med Assoc 53(4): 313–320

Yoboue CA, Hosch S, Donfack OT, Guirou EA, Nlavo BM, Ayekaba MO, Guerra C, Phiri WP, Garcia GA, Schindler T, Daubenberger CA (2022) Characterising co-infections with *Plasmodium* spp., *Mansonella perstans* or *Loa loa* in asymptomatic children, adults and elderly people living on Bioko Island using nucleic acids extracted from malaria rapid diagnostic tests. PLoS Negl Trop Dis 16(1):e9798

Zouré HGM, Wanji S, Noma M, Amazigo UV, Diggle PJ, Tekle AH, Remme JH (2011) The geographic distribution of *Loa loa* in Africa: results of large-scale implementation of the Rapid Assessment Procedure for Loiasis (RAPLOA). PLoS Negl Trop Dis 5(6):e1210

The Role of Human Host and Parasite Genetics in the Outcome of Loiasis

Jean Paul Akue

Abstract

Field studies suggest that genetics is involved in the outcome of *Loa loa* infection. This association could be due to parasite or human genetic factors. Although studies of the relationship between *L. loa* and genetic factors are limited, data from other filarial infections suggest the involvement of human HLA genes, immunoglobulin allotypes, and the parasite clone. These studies have helped shed light on the way forward in research on *L. loa.*

Keywords

Loa loa · Diversity · Cluster · Genetic factors · Clinical expression

1 Introduction

The diversity in the clinical and parasitological expression of *Loa loa* infection is consistent with the hypothesis that infection is influenced by genetic factors. In other filarial infections, genetic influences have been reported in field studies and/or animal models. Although similar studies have not been conducted for *L. loa,* the similarities between some species causing lymphatic filariasis and *L. loa* support a role for genetic factors in the outcome of loiasis. In the absence of sufficient data, we will compare what is known about *L. loa* with what has been reported on other filarial parasites with the aim of identifying the gaps in knowledge that need to be filled. In general, it has been demonstrated that the genetics of both the host and the parasite may play a role in the clinical expression of the disease, especially in experimental models of filariasis. It should be noted that results may differ

J. P. Akue (✉)
Neglected Parasitosis, CIRMF, Franceville, Gabon

J. P. Akue (ed.), *Loa loa: Latest Advances in Loiasis Research*,
https://doi.org/10.1007/978-3-031-49450-5_3

depending on the specific model used. For example, no differences were identified in susceptibility to infection by subperiodic *Brugia malayi* between male and female cats, while male *Meriones unguiculatus* were found to be more susceptible than females to infection with *Brugia pahangi* (Adam Ewert 1976; Lawrence R. Ash 1971). In cat studies with *Brugia pahangi* (Denham 1974), there was no difference in susceptibility between male and female animals, but female cats had a lower density of microfilariae than male cats: The arithmetic mean number of microfilariae in male cats was 4.27/mm^3 compared to 2.52/mm^3 in female cats ($p = 0.025$). Geometric mean microfilarial levels showed the same trend (3.06 vs. 1.9 microfilariae/mm^3, $p = 0.04$). Despite these conflicting results in animal models of filariasis, there are some observations from field trials in humans that implicate genetic factors in the clinical manifestations of filarial infections, which include human genetics, on the one hand, and parasite genetics, on the other hand. In this chapter, we dissect some of the most important genetic factors involved in loiasis and other filarial infections.

2 The Effect of Human Host Genetics in the Outcome of Loiasis

A striking observation in loiasis is that, despite exposure in an endemic area, only 30% of individuals become microfilaremic (Dupont et al. 1988). This assertion is backed by the fact that in areas of high density of infective bites, it was reported that only 37.6% of individuals became microfilaremic, with a mean level of 18.148 microfilariae per milliliter of blood. People who become microfilaremic often remain microfilaremic for life (Van Hoegaerden et al. 1987; Noireau et al. 1989; Garcia et al. 1995). This long-term stability seems to be associated with human host genetic factors. Another important observation is the fact that a study attempting to determine whether this trend is related to gender, age, or exposure to infective bites revealed that microfilaremia was gender-dependent, with males likely to be microfilaremic compared with their female counterparts (Pion et al. 2005). A segregation analysis of 74 nuclear families exposed to homogeneous loiasis transmission showed that there is a genetic predisposition involved, which explains the fact that the prevalence generally does not exceed 30% (Garcia et al. 1999). Another study of familial aggregation based on pedigree construct from 1126 individuals (Eyebe et al. 2018) found there was a familial tendency to be microfilaremic. Hypermicrofilaremia cluster and heritability were also higher between mothers and daughters. It appears that the likelihood for offspring from microfilaremic mothers to also be microfilaremic was high, but this was not the case for fathers (Eyebe et al. 2018). *Loa loa* treatment with current drugs induces fatal side effects; in this case, it is likely that human host genetics is implicated in the mechanism of side effects during treatment with drugs such as ivermectin or diethylcarbamazine (DEC). Indeed, this hypothesis is substantiated by experiments carried out with ivermectin in knock-out mice and dogs. These studies showed that the mutation of the multidrug resistance (*MDR1*) gene, which enables the synthesis of glycoprotein-P, can lead to coma through alterations in the blood–brain barrier (Mealy et al. 2001, 2002). This state resembles the one found during the treatment of *L. loa* hypermicrofilaremia. However, this hypothesis was tested in four individuals from Cameroon who experienced a post-IVM serious adverse event (SAE) and in nine non-SAE matched controls. No loss-of-function mutation was detected in *mdr-1* in any individual. However, haplotypes, associated with altered drug disposition, were present as homozygotes in two of the SAE patients (50%), but absent as homozygotes in the controls (0%) (Bourguinat et al. 2010).

Host genetics may also play a role in the difference in the clinical expression of an immune response to the disease. This could be driven by genetic diversity in the human leukocyte antigen (HLA) system or immunoglobulin (Ig) allotype polymorphisms. The genes coding the expression of these allotypes may show mutations with single nucleotide polymorphisms (SNP). Although these mutations have not been demonstrated in *L. loa*, their existence is possible as such phenomena have been shown in other human filarial infections; for example, in *Wuchereria bancrofti* numerous HLA alleles have been implicated in disease susceptibility. These include the allele B15 (Chan et al. 1984), allele B5 (Romia et al. 1988), or allele HLA-DQ5. HLA-DQ5-positive individuals have elevated IgG3 levels resulting in disease resistance (Yazdanbakhsh et al. 1995). In onchocerciasis, the DQ allele is associated with cutaneous manifestations (Brattig et al. 1986), while alleles DOB1* 0201 and DPB1 *0402 are associated with generalized disease (Meyer et al. 1994). In the same population, it was also noted that alleles DQA1* 0501 and DQB1* 0301 are associated with disease protection. Similarly, allotypes of immunoglobulin may affect the outcome of a filarial infection. The Igκ allotype (KM1) affects susceptibility in onchocerciasis, while KM3 leads to protection (Pandey et al. 1995). Similarly, the difference in a GM allotype affects the susceptibility to *Wuchereria bancrofti* (Kron et al. 1999). Another example is the appearance of an SNP in the gene coding the production of the inflammatory cytokine interleukin (IL)-13, which results in an amino acid change from arginine to glycine in position 110. Thus, this mutation results in a clinical expression known as localized and asymmetrical dermatitis (SOWDA) in onchocerciasis (Hoerauf et al. 2002). Genetic factors may also have an impact on innate immunity. For example, a study has identified both the genes of phagocyte-specific chitotriosidase (CHIT1) and of mannose-binding lectin (MBL) as a correlate to the susceptibility to *Wuchereria bancrofti* filaria (Choi et al. 2001). However, similar studies of loiasis have not been conducted to date. Thus, current knowledge on the implication of host genetics in loiasis needs to be expanded further. Genetic involvement seems plausible in the light of reports from a field study that reported the association between the outcome of *L. loa* infection and a genetic predisposition to microfilaremia. This observation is based on the segregation analysis of 74 nuclear families exposed to homogeneous loiasis transmission. The study demonstrated the existence of a dominant major gene that predisposed to microfilaremia (Garcia et al. 1999). This gene seems to be linked to the mother, as the probability of a microfilaremic mother, but not father, to deliver microfilaremic offspring was high. The gene implicated in such mechanism in loiasis is still unknown.

3 Genetic Heterogeneity in *Loa loa* Parasites

The genetic diversity of the *L. loa* parasite has not been studied as extensively as other filarial parasites, for example, *Onchocerca volvulus*. Some preliminary studies have been conducted (Higazi et al. 2004) using the internal transcribed spacer (ITS) of ribosomal RNA, the mitochondrial 16 rRNA gene, the ITS2 domain of the nuclear rRNA gene cluster, and the 15r3 polyprotein gene of *L. loa*. The work included samples from Nigeria, Gabon, the Democratic Republic of the Congo, and Cameroon. However, it was clear that both *ITS2* and *15r3* genes harbored polymorphisms. This suggests there is some genetic diversity among *L. loa* parasites. However, in this analysis, the sample size was small, not allowing definite conclusions to be drawn. Nevertheless, the findings of this study are strengthened by a recent work with MALDI-TOF analysis (Eyang Assengone 2021) using eight different samples from a neighborhood of Franceville, Gabon. The dendrogram shown in Fig. 1 stems from this analysis and presents two different branches for *L. loa*, which suggests the existence of two different strains. Whether these branches or clones are related to the clinical expression of loiasis is not known. Further studies are needed to analyze the relationship between these different clones and the diverse manifestations of *L. loa*, on an immunological, clinical, or parasitological level. As mentioned earlier, it is clear that studies on genetics and *L. loa* have just begun. We

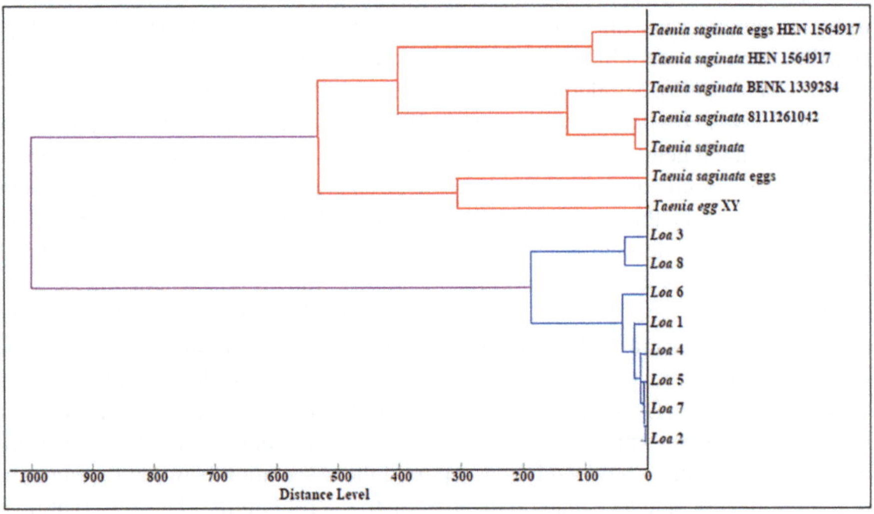

Fig. 1 Cluster analysis of *Loa loa* compared to *Taenia* species. MALDI–TOF MS profile similarities and distances were analyzed using hierarchical clustering of mass spectra of the eight samples of *Loa loa* microfilariae tested (derived from eight different individual donors), compared to *Taenia* species using the dendrogram function of MALDI Biotyper software v.3.3 (extracted from Eyang Assengone, these 2021).

will therefore focus on other filariae where more data are available. This will serve as a baseline to define what needs to be done in further studies of *L. loa* genetics.

Starting with the parasite *Wuchereria bancrofti,* genetic diversity has been demonstrated in several studies. Analysis of *Wuchereria* parasite populations in Pondicherry, India, and surrounding villages using random amplified polymorphic DNA (RAPD) showed a genetic heterogeneity of urban populations compared to the homogeneity of the rural population (Hoti et al. 2008). Similarly, in a posttreatment region of Papua New Guinea, sequence alignments of parasites were compared and genetic diversity in the *Wuchereria bancrofti* population was shown. This was done using the *Wuchereria bancrofti* mitochondrial cytochrome oxidase 1 (CO1) gene (Jongthawin et al. 2020; Small et al. 2013). In *Wuchereria bancrofti* also a change at position 200 of the beta-tubulin gene was shown and has been implicated in resistance against benzimidazoles (Churcher et al. 2008). However, the most well-characterized genetic influence on a filarial disease to date is the one described for *Onchocerca volvulus* (Dadzie et al. 1989; Zimmerman et al. 1992). Based on an amplification of the O-150 family (Erttmann et al. 1987, 1990) a repeated sequence specific to the genome of the parasite of the genus *Onchocerca* with strain-specific DNA probe pFS-1 (a forest strain) and PSS-IBT (a savannah strain), it was possible to relate some filarial genes with the clinical expression of the disease.

Thus, the use of these methods was able to predict the pathogenic potential of parasite populations throughout West Africa (Zimmerman et al. 1992). However, the relation between L. loa genetics and the clinical expression of *L. loa* disease is unknown with regard to encephalopathy that develops in some cases of loiasis, a report from Cameroon provides evidence that a specific strain has not been responsible for the encephalopathy observed in this area as there was no predominant strain of *L. loa* in the areas where SAEs were observed (Higazi et al. 2004). Based on this study, and on the immunological diversity observed in individuals exposed to or infected by *L. loa*, it is clear that research on the relationship between *L. loa* and genetics is needed.

4 Conclusion

Evidence derived from field studies indicates the involvement of genetic factors of the host in *L. loa* disease outcome. Despite the evidence provided by field studies and the clonal diversity reported for *L. loa*, no genetic factor of *L. loa* has been found to be associated with the clinical or parasitological outcome. Apart from a study in Cameroon of only four people with the *MDR1* mutation, the impact of HLA involvement or Ig allotype has not been studied in loiasis. These are important aspects that need to be addressed in further studies on the involvement of genetics in loiasis.

Acknowledgement We thank Amy Klion and Luzia Veletzky for English corrections and fruitful comments. The work on *Loa loa* was supported by the International Center for Medical Research of Franceville (CIRMF). CIRMF is sponsored by Total Gabon and Gabonese State.

References

Ewert A (1976) The comparative susceptibility of male and female and of mature and immature cats to infection with sub-periodic *Brugia malayi* by Adam Ewert* (Received for publication May 24, 1976). Rev Biol Trop 24(2):261–266

Brattig NW, Tischendorf FW, Reifegerste S, Albiez EJ, Berger J (1986) Differences in the distribution of HLA antigens in localized and generalized form of onchocerciasis. Trop Med Parasitol 37(3):271–275

Bourguinat C, Kamgno J, Boussinesq M, Mackenzie CD, Prichard RK, Geary TG (2010) Analysis of the mdr-1 gene in patients co-infected with *Onchocerca volvulus* and *Loa loa* who experienced a post-ivermectin serious adverse event. Am J Trop Med Hyg 83(1):28–32. https://doi.org/10.4269/ajtmh.2010.09-0734

Chan SH, Dissanayake S, Mak JW, Ismail MM, Wee GB, Srinivasan N, Soo BH, Zaman V (1984) HLA and filariasis in Sri Lankans and Indians. Southeast Asian J Trop Med Public Health 15(3): 281–286

Churcher TS, Schwab AE, Prichard RK, Basanez M-G (2008) An analysis of genetic diversity and inbreeding in *Wuchereria bancrofti*: implications for the spread and detection of drug resistance. PLoS Negl Trop Dis 2(4):e211. https://doi.org/10.1371/journal.pntd.0000211

Choi EH, Zimmerman PA, Foster CB et al (2001) Genetic polymorphisms in molecules of innate immunity and susceptibility to infection with *Wuchereria bancrofti* in South India. Genes Immun 2:248–253

Denham DA (1974) Studies with *Brugia pahangi*. 6. The susceptibility of male and female cats to infection. J Parasitol 60:642

Dupont A, Zue-N'dong J, Pinder M (1988) Common occurrence of amicrofilaraemic *Loa loa* filariasis within the endemic region. Trans R Soc Trop Med Hyg 82(5):730. https://doi.org/10.1016/0035-9203(88)90218-0

Dadzie KY, Remme J, Rolland A, Thylefors B (1989) Ocular Onchocerciasis and intensity of infection in the community II. West African rainforest foci of the Vector Simulium yahense. Trop Med Parasitol 40:348–354

Eyang Assengone Elsa Rush (2021) Caractérisation moléculaire et immunologique des antigènes de la microfilaire *Loa loa.* Thèse doctorat EDR 2021, PP 170

Eyebe S, Sabbagh A, Pion SD, Nana-Djeunga HC, Kamgno J, Boussinesq M, Chesnais CB (2018) Familial aggregation and heritability of *Loa loa*. Microfilaremia Clin Infect Dis 66(5):751–757. https://doi.org/10.1093/cid/cix877

Erttmann KD, Meredith SEO, Greene BM, Unnasch TR (1990) Isolation and characterization of form specific DNA sequences of 0. volvulus. Acta Leidensia 59:253–260

Erttmann KD, Unnasch TR, Greene BM, Albiez EJ, Boateng J, Denke AM, Ferraroni JJ, Karam M, Schulz-key H, Williams PN (1987) A DNA sequence specific for forest form Onchocerca volvulus. Nature 327:415–417

Garcia A, Abel L, Michel C, Ranque S, Richard P, Michel B, Jean-Philippe C (1995) Longitudinal survey of *Loa loa* filariasis in Southern Cameroon: long-term stability and factors influencing individual microfilarial status. Am J Trop Med Hyg 52(4):370–375. ISSN 0002-9637

Garcia A, Abel L, Cot M et al (1999) Genetic epidemiology of host predisposition microfilaraemia in human loiasis. Trop Med Int Health 4:565–574

Hoerauf A, Kruse S, Brattig NW, Heinzmann A, Mueller-Myhsok B, Deichmann KA (2002) The variant Arg110Gln of human IL-13 is associated with an immunologically hyper-reactive form of onchocerciasis (sowda). Microb Infect 4:37–42

Hoti SL, Thangadurai R, Dhamodharan R, Das PK (2008) Genetic heterogeneity of Wuchereria bancrofti populations at spatially hierarchical levels in Pondicherry and surrounding areas, south India. Infect Genet Evol 8(5):644–652. https://doi.org/10.1016/j.meegid.2008.06.002. Epub 2008 Jun 26

Higazi TB, Klion AD, Boussinesq M et al (2004) Genetic heterogeneity in *Loa loa* parasites from southern Cameroon: a preliminary study. Filaria J 3:4. https://doi.org/10.1186/1475-2883-3-4

Jongthawin J, Intapan P, Thanchomnang T, Sadaow L, Laymanivong S, Maleewong W, Sanpool O (2020) Morphological and genetic variation of *Wuchereria bancrofti* microfilariae in carriers in Thailand, Lao PDR and Myanmar: Evaluation using Giemsa-stained thick blood films. J Helminthol 94:E95. https://doi.org/10.1017/S0022149X19000865

Kron MA, Ramirez B, Belizario V Jr, Pandey JP (1999) Immunoglobulin allotypes among the Bicolanos of Sorsogon province, Luzon, Philippines: implications of phenotypes for filariasis. Exp Clin Immunogenet 16:65–71

Lawrence R. Ash (1971) Preferential susceptibility of male jirds (Meriones unguiculatus) to infection with *Brugia pahangi*. J Parasitol 57(4):777–780

Mealey KL, Bentjen SA, Waiting DK (2002) Frequency of the mutant MDR1 allele associated with ivermectin sensitivity in a sample population of collies from the northwestern United States. Am J Vet Res 63(4):479–481. https://doi.org/10.2460/ajvr.2002.63.479

Mealey KL, Bentjen SA, Gay JM, Cantor GH (2001) Ivermectin sensitivity in collies is associated with a deletion mutation of the mdr1 gene. Pharmacogenetics 11(8):727–733. https://doi.org/10.1097/00008571-200111000-00012

Meyer CG, Gallin M, Erttmann KD, Brattig N, Schnittger L, Gelhaus A, Tannich E, Begovich AB, Erlich HA, Horstmann RD (1994) HLA-D alleles associated with generalized disease, localized disease, and putative immunity in *Onchocerca volvulus* infection. Proc Natl Acad Sci U S A 91(16):7515–7519. https://doi.org/10.1073/pnas.91.16.7515

Noireau F, Carme B, Apembet JD, Gouteux JP (1989) *Loa loa* and *Mansonella perstans* filariasis in the Chaillu mountains, Congo: parasitological prevalence. Trans R Soc Trop Med Hyg 83(4): 529–534. https://doi.org/10.1016/0035-9203(89)90280-0

Pandey JP, Elson LH, Sutherland SE, Guderian RH, Araujo E, Nutman TB (1995) Immunoglobulin κ chain allotypes (KM) in onchocerciasis. J Clin Invest 96:2732–2734

Pion SD, Demanou M, Oudin B, Boussinesq M (2005) Loiasis: the individual factors associated with the presence of microfilaraemia. Ann Trop Med Parasitol 99:491–500

Romia SA, El-Ganayni GA, Makhlouf LM, Handousa AE (1988) HLA antigens and bloodgroups in bancroftian filariasis. J Egypt Soc Parasitol 18:211–220

Small ST, Ramesh A, Bun K, Reimer L, Thomsen E et al (2013) Population genetics of the filarial worm Wuchereria bancrofti in a post-treatment region of Papua New Guinea: insights into diversity and life history. PLoS Negl Trop Dis 7(7):e2308. https://doi.org/10.1371/journal.pntd.0002308

Van Hoegaerden M, Chabaud B, Akue JP, Ivanoff B (1987) Filariasis due to *Loa loa* and *Mansonella perstans*: distribution in the region of Okondja, Haut-Ogooué Province, Gabon, with parasitological and serological follow-up over one year. Trans R Soc Trop Med Hyg 81(3): 441–446. https://doi.org/10.1016/0035-9203(87)90163-5

Yazdanbakhsh M, Sartono E, Kruize YC et al (1995) HLA and elephantiasis in lymphatic filariasis. Hum Immunol 44:58–61

Zimmerman PA, Dadzie KY, DeSole G, Remme J, Soumbey AE, Unnasch TR (1992) *Onchocerca volvulus* DNA probe classification correlates with epidemiological patterns of blindness. J Infect Dis 165:964–968

Loiasis Disease Typical and Atypical Clinical Manifestations, Burden, and Local Aspects of the Disease

Luzia Veletzky and Wolfram G. Metzger

Abstract

Loiasis is a multifaceted disease with numerous signs and symptoms. The typical symptoms of "eyeworm" (subconjunctival migration of a *Loa loa* adult worm) and Calabar swelling occur alongside a variety of non-specific symptoms. Adult filariae and microfilaremia are associated with different manifestations. Due to the parasites' mobility, all body compartments may be affected by acute or chronic inflammatory reactions. Non-specific symptoms, such as malaise, headache, arthralgia, or pruritus, were most likely underestimated in the past but are increasingly recognized as important aspects of disease-related morbidity. Furthermore, chronic loiasis can be associated with severe organ damage affecting the heart, kidneys, and spleen and may also cause central and peripheral neurological symptoms. The incidence of these disease manifestations in affected populations is not yet known. Recently, significant excess mortality associated with microfilaremia has been observed in endemic populations in Cameroon and Congo. The enormous morbidity associated with the disease was captured in the first estimates of disease burden based on data from Gabon. In hyperendemic regions, loiasis is well known among the local population. The visible subconjunctival migration of a *Loa loa* adult worm ("eyeworm" episode) and Calabar swellings are often treated with traditional methods, sometimes leading to serious complications.

L. Veletzky
Department of Medicine I, Division of Infectious Diseases and Tropical Medicine, Medical University of Vienna, Vienna, Austria
e-mail: luzia.veletzky@meduniwien.ac.at

W. G. Metzger (✉)
Present Address: Department of Internal Medicine IV, Division of Diabetology, Endocrinology and Nephrology, Eberhard-Karls-University Tuebingen, Tuebingen, Germany
e-mail: wolfram.metzger@uni-tuebingen.de

J. P. Akue (ed.), *Loa loa: Latest Advances in Loiasis Research*,
https://doi.org/10.1007/978-3-031-49450-5_4

Keywords

Loiasis disease · Occult loiasis · Microfilaremic loiasis · Eyeworm · Calabar swelling · Mortality · Disease burden

1 Disease Manifestations

Loiasis has been a popular topic for case reports since it was first described in Western literature more than 250 years ago. Since then, infection with the parasite *Loa loa* has been associated with the typical sign of "eyeworm", in which the adult filariae visibly migrate under the patient's conjunctiva. While this disease phenomenon, along with the so-called Calabar swellings, is well known in endemic areas, it is still exciting enough for reports in non-endemic regions after hundreds of years. However, the fact that loiasis patients in endemic areas complain of headache, itching, and fatigue, as well as develop chronic organ damage and die earlier, has long been neglected.

When blood from infected people is put into cell culture for examination, the view through the microscope is a breathtaking sight. The person's blood may be teeming with young worms—the patient's blood is full of nematode larvae. In the hours after noon, when the horsefly bites, far more than 10,000 larvae (microfilariae) can appear in 1 ml of blood of an infected person. It is not known how many larvae remain in the tissue or attach to the cell walls of the blood vessels, but even the larvae that are flushed into the peripheral blood can number more than 70 million and weigh at least a gram in total, which corresponds to the weight of a raisin or an almond (personal communication, A. Renz/W. Hoffmann, Eberhard Karls University of Tübingen). However, these larvae are not always found in the peripheral blood of infected individuals, and the factors that determine the extent of microfilaremia remain unclear. The mechanisms underlying the differences in detectable microfilaremia are incompletely understood, but several factors, such as host and parasite genetics, parasite fecundity, and immune responses, have been proposed (Boussinesq and Gardon 1997; Eyebe et al. 2018; Pion et al. 2004, 2005). Furthermore, little is known about the number of adult worms dwelling in the human host. More than 100 years ago, the corpse of an infected person was dissected and 34 adult *L. loa* worms were discovered in the superficial connective tissue and under the aponeurosis limb muscles and tendons, primarily in the arms (Penel R 1905). As no further studies of this kind have been carried out to date, it is not known whether this number is below or above the average number of worms in a person. Models of adult worm populations based on the number of peripheral microfilariae in epidemiological datasets suggest that the average number of adult *L. loa* worms in a person may be higher (personal communication, A. Renz/W. Hoffmann, Eberhard Karls University of Tübingen).

Loiasis disease shares some general features with other filarial and helminth infections, which are important for understanding the diversity of associated signs and symptoms. First of all, loiasis is a chronic infection, as adult filariae can survive for more than 15 years in the human host and patients living in endemic areas are likely to be re-infected many times throughout their lives. This is evidenced by the fact that high antibody titers are found in endemic regions even in early childhood (Akue et al. 1997; Djikeussi and Akue 2014).

The infection, however, probably plays a role for the immune system even before that, since the parasite has been found in the placenta and there are even reports of detectable microfilaremia in newborns (Mombo-Ngoma et al. 2015; Pinder 1988). At the same time, microfilaremic individuals have been found to harbor stable parasitemia over time and appear to remain infected and retain their microfilaremia until senescence. Thus, in endemic populations, loiasis can be an infection that affects people throughout their lives.

For other parasite infections, it is known that the associated symptoms and manifestations depend on the parasite stage and the body compartment in which the parasite stage is located. For example, this can be illustrated by the various disease manifestations of schistosomiasis that range from acute, potentially fatal immune reactions known as Katayama fever to chronic, mainly asymptomatic disease leading to chronic organ damage including liver fibrosis and corresponding complications. Similar manifestations and underlying mechanisms are considered for loiasis.

Adult parasites and larval stages also appear to be associated with different courses of disease (Pinder 1988; Veletzky et al. 2022). The various manifestations of loiasis are thought to be caused by the adult filariae themselves, by attached or released microfilariae, or by immune reactions to the pathogen, including chronic and severe eosinophilia (Herrick et al. 2015). Since the adult filariae of *L. loa* are motile and migrate through various body compartments, the associated manifestations of the parasites' wanderlust are correspondingly diverse.

However, microfilaremia seems to be a marker or perhaps even a driving factor for serious organ manifestations and increased mortality. Only recently has it been impressively demonstrated that highly microfilaremic individuals have higher mortality, possibly due to organ damage (Buell et al. 2019; Chesnais et al. 2017; Hemilembolo et al. 2023).

It should be remembered that a proportion of people infected with large numbers of worms—both adult and larval—have hardly any apparent clinical symptoms. This fact, together with the finding that primary infections tend to be more acute in non-endemic individuals, indicates a high degree of immune tolerance induced by the parasite.

In summary, loiasis in endemic populations is a chronic infection that causes a range of bothersome and severe symptoms, as well as relevant mortality. However, knowledge about clinical loiasis disease is still incomplete. In returning travelers or in persons after short-term stays in endemic countries, the infection usually leads to acute reactions associated with eosinophilia, fever, and general malaise (Klion et al. 1991; Nutman et al. 1986). The following chapter discusses the existing evidence and theories for the disease symptoms of loiasis, the associated morbidity and mortality, disease burden estimates, and local aspects of the disease. As the relevance of loiasis is increasingly recognized, hopefully more will be known in the coming years.

1.1 Typical Disease Signs

Two specific disease manifestations have long been known in loiasis: the symptom of "eyeworm" (subconjunctival migration of a *Loa loa* adult worm) and the so-called Calabar swellings. Both have been described in Western literature for hundreds of years and seem to be well known in affected populations (Mongin 1770; Takougang et al. 2007). Both signs are caused by migrating adult filariae.

1.1.1 The Subconjunctival Migration of a *Loa loa* Adult Worm (Symptom of "Eyeworm")

The visible subconjunctival migration of the *Loa loa* adult worm not only gives loiasis its name, often referred to as "African eyeworm disease", but is also an important diagnostic feature and forms the basis for large-scale epidemiological studies (Takougang et al. 2002). Although it is a common phenomenon, well known in endemic populations and affected individuals, it is still a popular finding for case reports, as it was when first described in Western literature in 1770 by Mongin (Metzger and Mordmüller 2014; Mongin 1770).

For unknown reasons, the adult filariae migrate through the subcutaneous tissue of the host and occasionally invade the space under the conjunctiva. The movements of the emerging filariae are felt by patients and may be associated with periorbital swelling (see Fig. 1). If the parasite continues its way and enters the subconjunctival space, it becomes clearly visible under the transparent conjunctiva (see Fig. 2).

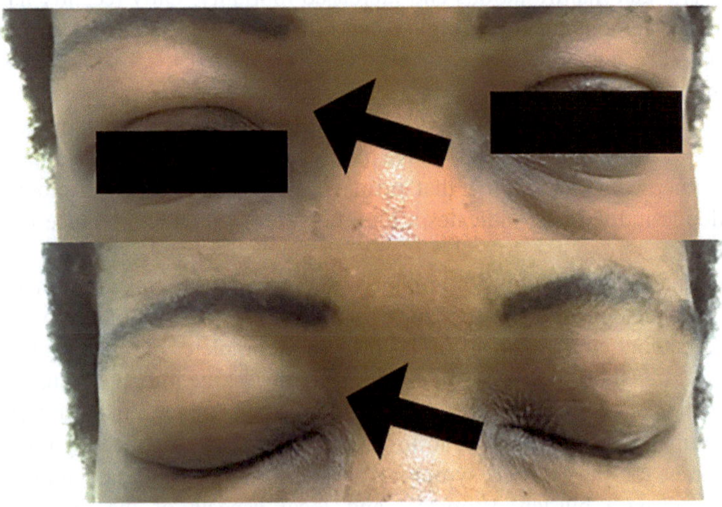

Fig. 1 An approaching worm may be felt by the patient and may cause periorbital swelling. Photos: L. Veletzky

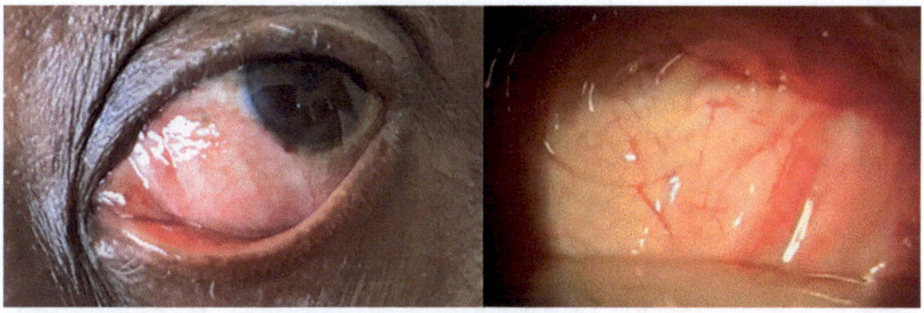

Fig. 2 Once the filaria enters the subconjunctival space, it may become visible (right side) and is often accompanied by conjunctival irritation and injections (left side). Photos: L. Veletzky

Without specific treatment, the filariae usually remain visible for a few hours to days before disappearing again into the tissue. Why adult *L. loa* filariae have a predilection to migrate under the conjunctiva, or whether this phenomenon occurs purely by chance, is unknown.

For the patient, the "eyeworm" episode can cause local irritation of the eye, sensitivity to light, and visual disturbances. In addition, the presence of the worm is understandably considered disgusting and repulsive for the individual and may trigger a feeling of shame (personal communication of patients to the authors). Affected persons report that the worm crossing the eye keeps them from work and their daily chores (Veletzky et al. 2020, 2021). The frequency of occurrence of "eyeworm" episodes varies greatly: Some people report several episodes per year; others report only one episode in their lifetime. People who reside only temporarily in endemic areas or travelers returning are less likely to be affected by worm migration through the eye. It is hypothesized that the difference in frequency might be due to varying degrees of adult worm burden (Noireau et al. 1990).

It is important to note that other filariae have also been described to reach the eye and become visible under the conjunctiva, including *Wuchereria bancrofti*, *Thelazia* spp., *Dirofilaria immitis,* and *D. repens*. As the endemicity of loiasis is restricted to parts of West and Central Africa, the diagnosis of loiasis due to an "eyeworm" episode without prior travel to endemic areas should always raise suspicion of other filarial diseases such as dirofilariasis.

1.1.2 Calabar Swelling

Calabar swellings are transient, non-pitting edema that have long been associated with loiasis (see Fig. 3). Named after a Nigerian port, these swellings are considered pathognomonic in patients who have lived in or visited an endemic area in the past. Calabar swellings are often described as occurring above the wrist but may occur elsewhere on the body. They can last from hours to days and may be described as itchy or painful, most likely depending on the location of the filariae, the associated swelling, and the involvement of the nerves in the local inflammation. Due to their presence above the wrists, they are described as hindering the performance of work as they prevent a firm grip. The exact mechanism leading to the development of Calabar swellings is not known, but it is thought to be caused by a local immune

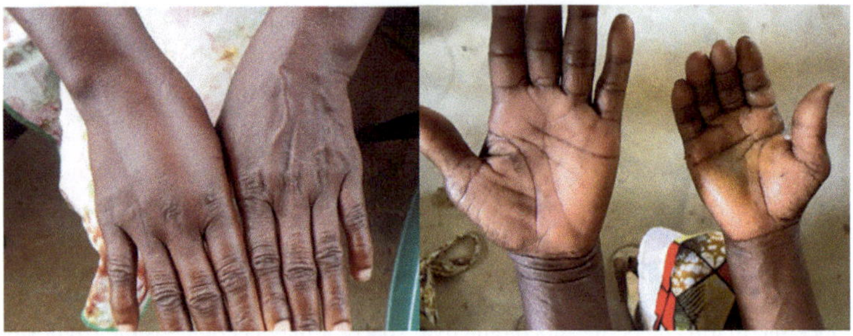

Fig. 3 Calabar swellings seen in patients from Gabon. Often described as occurring above the wrists as they affect other body parts as well. They might be associated with itch or pain and last for hours to days. Photos: D. Stelzl, L. Veletzky

response to moving adult filariae or to released microfilariae (Pinder 1988). While Calabar swelling is a well-known sign of loiasis, its reported prevalence is less correlated with the prevalence of microfilaremia in endemic populations compared with the symptom of "eyeworm" (Takougang et al. 2002). This is thought to be due to the limited specificity of the sign and most likely due to frequent misdiagnosis of other causes of joint swelling or overlooking Calabar swelling in atypical body regions (Takougang et al. 2002). Calabar swellings have been described to occur more frequently in returning travelers than in patients living in endemic regions (Klion et al. 1991; Nutman et al. 1986).

1.2 Unspecific Symptoms

For a long time, loiasis was not thought to cause relevant symptoms in endemic populations, apart from the well-known signs of the Calabar swelling and "eyeworm". However, as more and more data become available to assess disease manifestations, this assumption has been clearly refuted. Patients are considered to have a number of non-specific symptoms in addition to the "typical" signs. While there are limited data on subjective symptoms in loiasis patients living in endemic regions, several symptoms have repeatedly been associated with the infection. Associated non-specific symptoms include joint and muscle pain, headache, fatigue, asthenia, pruritus, urticaria, and soft tissue swellings in any body part (Bouyou-Akotet et al. 2014, 2016; Carme et al. 1989; Noireau et al. 1990; Veletzky et al. 2020).

Pruritus has been associated with loiasis infection, but the features described are variable and not all analyses have found a link with the infection (Akue et al. 2011; Noireau et al. 1990; Veletzky et al. 2020, 2022). The associated pruritus has been described as episodic, occurring in bouts, and sometimes centered on the upper extremities, but may also be generalized (i.e., involving the whole body) (Carme et al. 1989; Noireau et al. 1990). These contradictory results may be due to different methods of diagnosing loiasis and how pruritus is interrogated, as well as possible other causes of pruritus. Furthermore, the results of loiasis studies are often difficult to compare as the diagnosis of loiasis and the criteria for defining the disease are not standardized and therefore can differ between publications. In addition, non-specific symptoms may be multifactorial and polyparasitism may play a role in the development of symptoms.

Interestingly, in a population-based study, it was found that individuals with signs of migrating adult worms indicated by a positive history of "eyeworm" reported significantly more symptoms than individuals who had only microfilaremia (Veletzky et al. 2022). This included amicrofilaremic individuals who only had a positive history of "eyeworm", a condition often referred to as "occult loiasis". So, despite the increased risk of developing severe organ manifestations, people with high microfilaremia seem to have fewer subjective complaints—at least in the study population described. One hypothesis is that in amicrofilaremic but parasitized individuals, the immune system manages to suppress the release of microfilariae or neutralize the released larvae. This immune activation can cause non-specific and general symptoms such as malaise, fatigue, or arthralgia (Pinder 1988). However, in microfilaremic individuals, the immune system may be less active, whether due to immunomodulatory, parasite-dependent, or genetic host factors, so that microfilariae are not neutralized and thus do not cause generalized symptoms.

The subjective well-being of microfilaremic individuals may also have played a role in the fact that loiasis was often described as asymptomatic in endemic populations in the past. It is noteworthy that traditionally loiasis was often diagnosed only by the detection of microfilaremia and not by the history of "eyeworm", until the importance of occult loiasis was recognized (Dupont et al. 1988; Pinder 1988).

1.3 Eosinophilia

Eosinophils are important immune effector cells for host defense against helminth infections, and eosinophilia may be caused by many different helminthic pathogens (see Fig. 4). However, their exact effects and roles in the defense against acute and chronic helminth infections, as well as their role in the development of pathologies associated with chronic helminth infections, are not yet fully understood. It is known that the effect of eosinophils depends on their activation, which can be assessed by eosinophil activation markers and their recruitment into the respective tissue. Importantly, acute or chronic eosinophilia due to eosinophilic inflammation can cause organ damage (Andy et al. 1998; Mitre and Klion 2021; Rosenberg et al. 2013).

Generally, loiasis is associated with eosinophilia, defined by an absolute eosinophil count above 500/μl blood. Importantly, loiasis has been associated with hypereosinophilia reaching extremely high levels of eosinophils—a peculiarity of the disease (Nutman et al. 1986). This has been demonstrated in people living in endemic regions and in returning travelers and short-term residents. Eosinophilia has been repeatedly described in case reports of people with atypical disease manifestations of loiasis and is often more pronounced in primary infections, e.g., in people who have not lived in endemic regions for a long period of time (Bouchaud

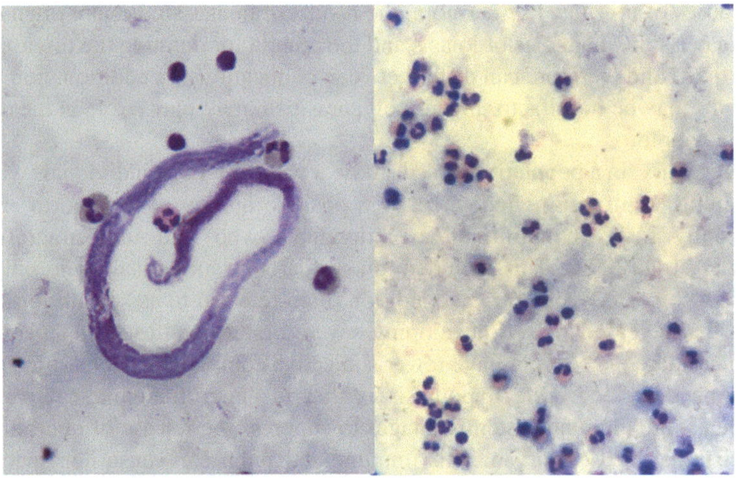

Fig. 4 A Giemsa-stained *Loa loa* microfilaria surrounded by basophils and an eosinophil at 1000× magnification with oil (left side); a Giemsa-stained thick blood smear showing extensive eosinophilia of about 8000 eosinophils/µl at 400× magnification, seen in a returning traveler (right side). Photo: L. Veletzky

et al. 2020; Carme et al. 1989; Churchill et al. 1996; da Silva et al. 2015; Gantois et al. 2013; Klion et al. 1991; Pinder 1988). Furthermore, several studies indicate significant differences in cell activation between infected people living in endemic areas and infected returnees or short-term residents as measured by cytokine levels relevant for eosinophil activation (Herrick et al. 2015; Klion et al. 1991; Nutman et al. 1986).

Nevertheless, eosinophilia might also be high in individuals living in an endemic region as shown in an analysis of 1232 individuals from an endemic region. While it was seen that loiasis was generally associated with eosinophilia, the extent of eosinophilia was also found to vary between the different types of infection, i.e., microfilaremic and occult loiasis. In this analysis, moderate eosinophilia was found in "eyeworm"-positive individuals, but the levels of the effector cells were higher in microfilaremic individuals. In addition, the extent of eosinophilia was related to the extent of microfilaremia, and none of the 56 subjects with more than 8000 microfilariae per ml of blood had an eosinophil count of less than 500/µl (Veletzky et al. 2022).

These findings contrast with another study, where data from 204 individuals seen and treated in the USA were analyzed. The patient group comprised returning travelers and migrants from endemic regions. Of the 204 individuals included, 70 were microfilaremic and 134 amicrofilaremic. Here, higher eosinophilia, antifilarial antibody, and interleukin levels were found in amicrofilaremic individuals compared with those being microfilaremic (Herrick et al. 2020).

One of the reasons for these differences might be the different compositions of patient groups, i.e., patients living in endemic regions and returning travelers and time since their last exposure, frequency of recurrent infections, or co-infections with other parasites.

While more studies are needed to understand the exact role of eosinophils in loiasis, these cells seem to have a major role in disease pathology.

1.4 Atypical and Serious Organ Manifestations

Loiasis disease has long been associated with complications of the cardiovascular system, kidneys, lungs, and central nervous system. However, there are only few data on occurrence and frequency of these serious disease outcomes. Due to the specific biology of *L. loa*, a parasite with high mobility in the host, the filariae and microfilariae may reach all parts of the body. Various severe but atypical manifestations have been described for all different body compartments and organs, resulting in a quite heterogeneous clinical picture. This could be one of the reasons why loiasis-related manifestations may be overlooked or misclassified. While the typical symptoms of "eyeworm" and Calabar swelling are caused by adult worms, atypical disease manifestations may be related to the presence of adult filariae, microfilariae, or both.

In 2019, Buell et al. published a systematic review of atypical disease manifestations of loiasis, summarizing the extensive clinical information available from case reports (Buell et al. 2019). Case reports are still the main source of clinical information on loiasis, and it was found that almost half of the patients described had atypical disease manifestations. Of course, it has to be assumed that this high frequency is due to publication bias, but it nevertheless shows that these manifestations are not rare and do occur. It must be emphasized that the involvement of almost all organ systems was described and that the manifestations of loiasis were quite diverse. Manifestations described included damage or pain due to the physical presence of living or calcified adult worms, acute tissue inflammation with or without effusions in various compartments, or chronic tissue damage such as fibrosis.

Effusions were found in the lungs, joints, or intra-abdominally as ascites. In the described cases where fluid was collected and examined, microfilariae could be detected. In those cases where a differential blood count was obtained, eosinophilic effusions were detected. So, it seems that effusions caused by loiasis are more likely to be eosinophilic and contain microfilariae. This is consistent with descriptions of *L. loa* microfilariae being found in all types of body fluids, including urine, ascites, cerebrospinal, and follicular fluid. While a blood contamination of the fluids as a reason for the presence of microfilaria is difficult to exclude with full certainty, the consistent description of the finding in different body fluids supports its relevance (see Fig. 5) (Pinder 1988; Puy et al. 2017).

Cardiac Involvement

One of the most serious organ complications associated with loiasis is endomyocardial fibrosis, leading to heart failure. Endomyocardial fibrosis in

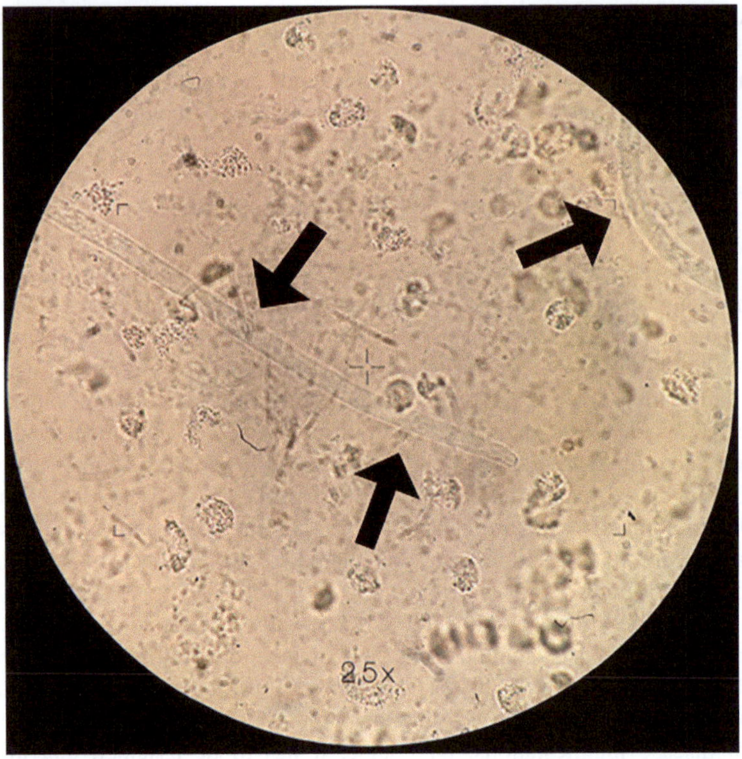

Fig. 5 *Loa loa* microfilariae have been found in various body fluids and can be identified by microscopic examination of samples. The microfilariae shown here were found in a blood sample that had been frozen for over a year and thawed for further laboratory testing. The sample was examined under a light microscope without staining at 1000× magnification using a cover slip and oil. Without staining, the parasite sheath may not be visible. Photo: L. Veletzky

hypereosinophilia is most likely caused by chronic tissue remodeling due to persistent tissue inflammation. This has been supported by the detection of activated eosinophils and major basic protein in histological myocardial samples from patients with eosinophilic endomyocardial disease (Tai et al. 1987). A similar mechanism is hypothesized to be the underlying cause in the case of endomyocardial fibrosis in loiasis as endomyocardial fibrosis has been described in several case reports in young people with eosinophilia and detectable *L. loa* microfilaremia living in endemic areas (Andy et al. 1981, 1998; Berenguer et al. 2003; Buell et al. 2019; León et al. 2012). It was also found that the prevalence of endomyocardial fibrosis is higher in the endemic regions of loiasis. However, a clear causal relationship between endomyocardial fibrosis and loiasis has not yet been established, and clinical data may provide further insight into the pathophysiology of this complication. There are no specific treatment recommendations in this regard, apart from the treatment of heart failure in case of functional impairment. Should *L. loa* microfilaremia be causative for this complication or be an important contributor to it, it would of course be of great importance to have at least a microfilaricidal treatment, long before the functional remodeling of the tissue and the functional impairment begin.

Pulmonary Involvement

Several case reports have described respiratory symptoms associated with loiasis. Disease manifestations included acute development of pulmonary infiltrates or pleural effusions, but also pulmonary fibrosis (Buell et al. 2019; Cambanis 2010; Hulin et al. 1994; Klion et al. 1992). *L. loa* microfilariae were found in both pleural effusions and bronchoalveolar lavage fluid and were accompanied by eosinophilia. Treatment with appropriate anthelmintics led to a rapid improvement of symptoms.

Renal Involvement

Kidney damage has been repeatedly associated with loiasis, and there are several case reports describing this organ complication (Hall et al. 2001; Katner et al. 1984; Ngu and Youmbissi 1987; Pakasa et al. 1997; Pillay 1973). The extent of documented kidney damage varies widely, but it is unclear whether this is because the diagnosis is made at different stages of progressive disease or because the disease tends to cause damage of varying degrees. Immunocomplex-derived damage has been suggested as an important factor in loiasis-related kidney damage (Ngu et al. 1985). However, case reports indicate that kidney damage caused by loiasis is usually glomerular disease but described kidney damage ranged from proteinuria without symptoms or further complications to end-stage renal disease. In the cases where renal biopsies were performed, histopathology most commonly showed membranous glomerulopathy, but also focal interstitial inflammation, focal segmental glomerulosclerosis, interstitial fibrosis, and glomerulosclerosis. Microfilariae were detected in the urine, so-called microfilaruria, or in the histopathological samples. This suggests direct involvement of microfilariae in kidney damage, but deposition of immune complexes has also been postulated as an underlying mechanism of kidney damage. Treatment also includes functional support for the kidneys and antiparasitic treatment. However, initial deterioration of renal function after microfilaricide treatment has been described in case reports.

Spleen Involvement

Involvement of the spleen in loiasis infection has long been postulated and seems reasonable due to the central role of the spleen. The spleen is one of the main driving forces of the immune system, covering various areas of pathogen defense, also against parasites, such as the removal of *Plasmodium*-infected erythrocytes from the blood. Hyposplenism, whether functional or anatomical, leads to an increased risk of severe infections. Examples include malaria, where higher parasitemia develops more rapidly and patients are at higher risk of severe malaria, or fulminant infections caused by encapsulated bacteria, leading to rapid disease progression and often patient death. Consistent with these known effects, it was observed in animal models of loiasis that monkeys developed higher microfilaremia after splenectomy. It was therefore hypothesized that a functioning spleen would intercept at least some of the circulating microfilariae, leading to stabilization of microfilaremia over time (Hawking et al. 1967).

In case reports, loiasis has been associated with the development of transient splenic nodules. In some cases, these spleen lesions in combination with hypereosinophilia led to the differential diagnosis of lymphoma and thus splenectomy was done. However, histopathology demonstrated eosinophilic granulomas with microfilariae (Burchard et al. 1996). Two cases of young men with splenic lesions and peripheral eosinophilia were followed over a period of about 2 years, showing that the splenic lesions regressed after filaricidal treatment (Tamarozzi et al. 2022). Splenic lesions were visualized by ultrasound and magnetic resonance imaging (MRI), which showed hypoechoic lesions that decreased in size and number until imaging was near normal. Importantly, both affected individuals had low or moderate peripheral microfilaremia at the time of diagnosis.

The role of the spleen and associated pathology of the spleen with loiasis, however, is still unclear, and more data are warranted.

Central and Peripheral Neurological Manifestations of Loiasis

Central and peripheral neurologic manifestations have long been described in loiasis patients. Associated neurological manifestations include peripheral neuropathy and psychological symptoms, as well as more serious outcomes such as treatment-related or spontaneous fatal encephalopathy. Non-specific neurological symptoms such as fatigue, asthenia, somnolence, sensory disturbances, and motor deficits have been described in *L. loa* infection, and severe headache has also been described (Bogaert et al. 1955; Buell et al. 2019). Furthermore, peripheral involvement in the form of paresthesia, peripheral nerve palsies, or temporary immobility of the extremities has been associated with the infection (Bhalla et al. 2013; Veletzky et al. 2020).

It is known that encephalopathy may occur after treatment with diethylcarbamazine (DEC) or ivermectin in patients harboring high microfilaria loads and that the risk is much higher in individuals harboring more than 30,000 mf/ml blood. Treatment-related serious adverse events (SAEs) of varying degrees are known and may occur in the days following treatment. SAEs have been described to start 24 h after treatment and may include fatigue and personality changes, including agitation or mutism, and extrapyramidal signs. In some cases, impaired consciousness occurred in the days following treatment and progressed to coma. Clinical examinations showed retinal hemorrhages and the presence of microfilariae in the CSF and urine. Electroencephalograms showed a diffuse pathology. While some patients recovered without sequelae, some remained impaired. Deaths occurred due to difficulties in access to adequate health care during coma, for example, due to exacerbation of underlying diabetes or development of bed sores (Boussinesq et al. 1998, 2003; Gardon et al. 1997). There are also case reports describing spontaneous encephalopathies presenting as coma without prior filaricidal treatment in highly microfilaremic individuals. Interestingly, in these cases, most patients spontaneously improved after a few days, although one patient died a few days later after having convulsions for unknown reasons and another died after receiving DEC treatment (Lukiana et al. 2006). In an animal model of highly microfilaremic baboons which were treated with ivermectin, eosinophilic and macrophage infiltration of the brain tissue was seen. These changes causing microvascular damage were associated with neurological findings, and similar mechanisms were suggested as cause for the neurological adverse events after treatments in humans (Wanji et al. 2017).

In 2011, Gobbi et al. described the case of a woman with detectable *L. loa* microfilaremia and a sensorimotor deficit involving the ulnar nerve, which manifested as typical clawing of the hand. In this case, an electromyogram showed severe abnormalities. After treatment of the loiasis, her symptoms resolved and she regained mobility of her fingers (Gobbi et al. 2011). Similar cases have been repeatedly described, and one hypothesis is that adult filariae may cause localized symptoms through mechanical interference and subsequent local inflammation of nerves (Bhalla et al. 2013; Bogaert et al. 1955; Buell et al. 2019). The presence of co-infections with malaria or bacteria has been hypothesized to weaken the blood-brain barrier, allowing *L. loa* microfilariae to enter the CSF (Bogaert et al. 1955; Lukiana et al. 2006). After treatment with ivermectin, microfilariae could be detected in brain tissue.

Atypical Eye Involvement

Apart from the classic "eyeworm", in which the adult filariae migrate under the conjunctiva, loiasis has been associated with various forms of ocular disease. In some cases, it was possible to identify the parasite in atypical locations in the eye. Both adult filariae and microfilariae were found in the anterior and posterior chambers (Beaver 1989; Carme et al. 1984; Hassan et al. 2016; Védy et al. 1975). This atypical localization of the parasite was accompanied by visual disturbances and often loss of vision. However, detection of the parasite in ophthalmic compartments was not possible in all cases, and the findings described covered a wide spectrum and included uveitis, chorioretinitis, corneal edema, posterior chamber opacification, vitreous hemorrhage, hemorrhagic retinopathy, retinal detachment, neovascularization, and retinal artery occlusion (Buell et al. 2019).

Gastrointestinal Involvement

Microfilariae have been isolated from ascites, peritoneal and gastric lavage (Buell et al. 2019; Hautekeete et al. 1989; Whitaker et al. 1980). In one case, involvement of the bowel was described, presenting as intestinal obstruction (Neafie et al. 1985). Histopathologic examination of biopsies from the constructing lesion revealed fibrosis and microfilariae.

Joint Involvement

While Calabar swellings tend to occur over or next to joints, loiasis has also been described to cause other types of joint involvement. There are several cases in the literature where patients presented with acute septic or aseptic arthritis, and both microfilariae and eosinophilia were detected in aspirated synovial fluid, suggesting that the parasite was indeed the cause of the acute inflammation (Jaffres et al. 1983; Roussel et al. 1989).

Reproductive Organs
Adult filariae have been identified as the cause of testicular swelling, and filariae have been surgically collected from the tunica vaginalis and spermatid cord. Microfilariae were detected in cervical smears and follicular fluid from women who underwent in vitro fertilization. Whether the presence of microfilariae has an effect on fertility is unclear (Chang et al. 2008; Puy et al. 2017; Wisanto et al. 1993). Microfilariae of *L. loa* have also been found in a fluid derived from hydrocele fluid after a surgical intervention on male testis (personal communication, JP Akue).

Lesions Caused by Calcified Filariae
Dead and calcified filariae have been repeatedly identified as the cause of painful and clinically silent lesions. Several cases have been published in which likely calcified adult filariae were identified in mammograms. These calcifications in breast tissue were asymptomatic or caused painful lumps and were described as elongated structures on imaging (Carme et al. 1990). Similarly, calcified filarial structures have been detected on radiographs of painful joints, most likely causing pain and dysfunction by mechanical irritation (Williams 1954). Based on the understanding of the motility of adult *L. loa* filariae, their ability to reach different body compartments, and their long but still limited life span, such calcified lesions could occur at different body sites. Because it has been noted that they are not necessarily accompanied by symptoms, such calcified lesions of decaying adult filariae may be more common than is known.

Skin Involvement
In addition to pruritus, urticarial lesions are also associated with loiasis. The adult worm can become visible and palpable migrating under the outer layers of the skin (see Fig. 6). In a study using PCR for species diagnosis, it was found that *L. loa* microfilariae may be present in skin sections prepared for diagnosis of onchocerciasis. Because skin snips are considered highly specific for the latter, *L. loa* skin infections may have been misdiagnosed and the associated symptoms underestimated (Nana-Djeunga et al. 2019).

2 Disease Burden of Loiasis

In recent years, new insights have been gained into the morbidity, mortality, and disease burden associated with loiasis. Chesnais and co-authors first demonstrated in 2017 that the disease is associated with increased mortality in endemic populations in Cameroon, and this was recently confirmed by a second study in Congo (Chesnais et al. 2017; Hemilembolo et al. 2023). In 2020, the first burden of disease estimate based on morbidity data from Gabon was published (Veletzky et al. 2020). Both efforts were important steps toward recognition of loiasis as a relevant disease, but many factors contributing to disease burden remain unclear. Initial disease-specific DALYs calculated based on morbidity caused in an endemic region were similar to recognized NTDs such as schistosomiasis (Veletzky et al. 2020). Of note, this first

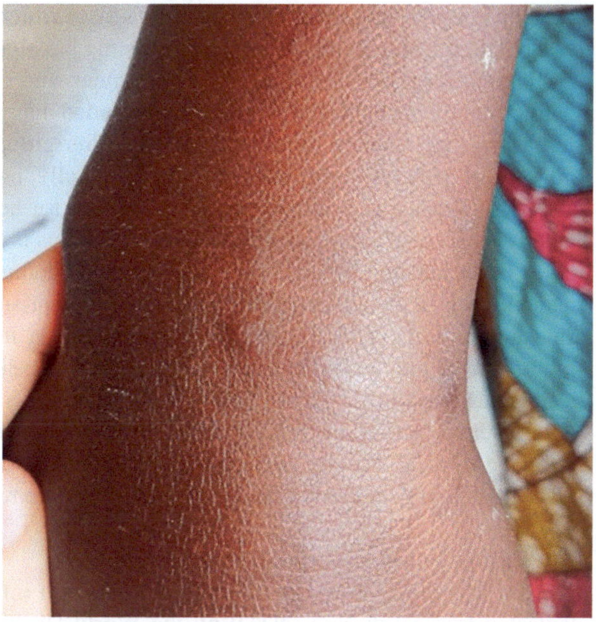

Fig. 6 An adult worm was palpated by the patient under the skin of the right wrist. She reported that the worm-like structure appeared from time to time under the skin in different parts of the body, moving about 1–2 cm/day and often becoming visible, as seen in the picture. She did not have Calabar swelling at the time the picture was taken. Photo: L. Veletzky

DALY estimate was based solely on common, disabling but not life-threatening symptoms and did not account for organ damage, rare acute manifestations (such as spontaneous encephalopathy), and mortality associated with loiasis. Therefore, the actual burden of disease, including all consequences, is likely to be much higher—for the affected individual, but also for the entire endemic population.

Loiasis, Disease Burden, Neglected Tropical Diseases, and Sustainable Development Goals

Burden of disease studies aim to objectively quantify the impact of disease on the affected population and provide the basis for health policymakers in allocating resources to best improve the health of the population. To this end, Global Burden of Disease (GBD) studies have been conducted since the 1990s.

Disability-adjusted life years, in short DALYs, are the main units used for measuring disease in these GBD studies. DALYs are based on two main principles: first, the principle that "like goes with like", and second, the principle that no individual non-health aspects such as wealth or socioeconomic status are taken into account when calculating health loss in a population, everyone's health loss is weighted equally, regardless of socioeconomic status (Murray and Acharya 1997). For DALY calculations, symptoms and health status are graded according to severity and caused disability and assigned to so-called disability weights, regardless of the underlying cause. Neglected tropical diseases (NTDs) are diseases associated with poverty and are not given sufficient attention by researchers in developing diagnoses, treatments, or prophylaxis, or by policymakers in public health interventions, compared with their health impact on the global population. NTDs are defined by WHO and include a list of diseases which are prioritized by WHO for research, treatment, and eradication. This listing has significant implications, including funding for public health research, treatment, and interventions.

Importantly, recent GBD studies have included many NTDs. This is an important step in raising awareness and attention to this group of diseases, which affect millions of patients worldwide but are generally not endemic in high-income countries. The use of health metrics that account for health losses regardless of the underlying disease or wealth of affected populations is an attempt to equalize health losses in an unequal world. The United Nations defined the Sustainable Development Goals as a global development strategy, comprising health, development, and education aspects. These efforts have supported an increasing recognition of the importance of previously underestimated causes of health loss and NTDs in global health and have vastly improved global understanding of the impact of disease, led to public health action, and supported the development of programs to combat neglected tropical diseases, among others. However, to be included in international GBD studies or listed as NTDs, the diseases must be recognized as such.

Calculating Loiasis-Specific Disease Burden

Because loiasis has long been considered a benign infection and is not listed as a neglected tropical disease (NTD) by the World Health Organization, it has never been included in Global Burden of Disease (GBD) studies. A first estimate for loiasis-associated disease burden expressed as disability-adjusted life years (DALYs) was published in 2020 (Veletzky et al. 2020). The data used for the DALY calculation were collected as part of a cross-sectional survey and included a variety of symptoms associated with the disease (see Fig. 7). Individuals were also asked whether and how the symptoms affected them in their daily lives. Questions addressed the impact of symptoms on work, sleep, and psychological well-being of those affected. These are important factors in classifying the impact of symptoms. Importantly, associated mortality was not included in the DALY calculations, nor were any symptoms or severe disease manifestations associated with loiasis. This was done partly for lack of data on which to base the estimate, but also to avoid overestimation of disease burden in this first estimate.

Four main symptoms were used for the DALY calculation: "eyeworm" (subconjunctival migration of the *Loa loa* adult worm), Calabar swelling, severe headache, and arthralgia, which were found to be significantly associated with loiasis.

For DALY calculation, the incidence of symptoms in the loiasis-positive group and in the loiasis-negative group within the preceding year was calculated. Symptom incidence in the loiasis-negative group was then subtracted from the loiasis-positive group and stratified for the age groups commonly used in Global Burden of Disease studies to obtain the attributable proportion of symptom incidence. The excess of reported symptoms was counted as the attributable fraction and used for the DALY calculation.

Disability weights were taken from the 2015 GBD study. The available disability weights were chosen to best match the symptoms associated with loiasis. The factors

Fig. 7 Interviews for the loiasis disease burden study were conducted in participating villages in rural Gabon, often using the shade of large trees. Photo: L. Veletzky

considered in the weighting were related to the magnitude, duration, and psychological impact of symptoms based on data collected in the survey. Current population data, including the distribution of urban and rural populations, were taken from the 2017 World Population Prospects and 2013 Gabonese census data (Ntsame Ondo 2015; "World Urbanization Prospects – Population Division – United Nations" 2018). DALYs were calculated in absolute and relative terms per 100,000 population for rural and urban populations and for the total population of Gabon.

Because of the complexity of the disease and limited data on disease manifestations, estimates were conservative in the weights of disability attributed but also in the inclusion of symptoms. Nevertheless, DALYs caused by loiasis were found to be 412.9 (95% CI: 273.9–567.7) per 100,000 of the highly endemic rural population, which is comparable to other parasitic NTDs. While the calculated relative DALYs for the urban population were correspondingly lower due to the lower prevalence of the disease, the total DALYs caused for the entire country of Gabon were an estimated 1581.6 (95% CI: 1053.1–2176.9), which is immense for a population of less than 2 million people. Comprehensive DALY results can be found in Table 1. These figures show the extreme impact of the disease on endemic populations, although they are most likely grossly underestimated due to the use of very conservative settings. Importantly, loiasis prevalence is not evenly distributed within endemic regions and loiasis disproportionately affects rural populations because of the habitat and biology of the vector, differences in living

Table 1 DALY estimates are stratified by causative symptom, by region (rural vs urban), and for the entire Gabonese population. DALYs are reported in absolute numbers and relative per 100,000 populations. Estimated numbers of cases are reported as absolute values per region. The table was adapted from Veletzky et al. (2020)

Symptom-specific DALYs and affected cases by population				
	DALYs absolute		DALYs per 100,000	
Rural population (n = 204,474)				
Symptom				
"Eyeworm"	244.6 (61.7–497.5)	37,621	119.9 (30.0–242.9)	18,399
Calabar swelling	30.0 (5.8–63.0)	5055	14.8 (2.9–30.9)	2472
Arthralgia	477.5 (323.1–632.1)	3929	233.1 (157.8–309.0)	1922
Severe headache	91.9 (38.9–163.8)	4837	45.1 (18.8–80.3)	2366
Overall	843.2 (558.6–1163.2)	51,442	412.9 (273.9–567.7)	25,158
Urban population (n = 1,724,526)				
Symptom				
"Eyeworm"	210.2 (53.7–424.9)	32,121	12.2 (3.1–24.6)	1864
Calabar swelling	26.0 (5.0–55.1)	4391	1.5 (0.3–3.2)	255
Arthralgia	365 (245.6–482.6)	2997	21.1 (14.2–28.0)	174
Severe headache	74.4 (31.5–132.4)	3940	4.3 (1.8–7.7)	229
Overall	675.2 (442.2–939.6)	43,469	39.2 (25.6–54.5)	2521
Overall Gabon (n = 1,929,000)				
Symptom				
"Eyeworm"	448.4 (113.3–914.2)	69,050	23.4 (5.9–47.4)	3580
Calabar swelling	52.6 (10.2–110.4)	8820	2.7 (0.5–5.7)	457
Arthralgia	913.7 (617.2–1208.7)	7506	47.3 (32.0–62.6)	389
Severe headache	168.7 (71.2–302.4)	8937	8.8 (3.7–15.7)	463
Overall	1581.6 (1053.1–2176.9)	94,313	82.2 (54.5–113.2)	4889

habits, and thus differences in exposure. At the same time, the most affected populations have less access to health care, exacerbating inequality within countries. These are important factors in calculating the burden of disease and demonstrate the impact of loiasis on achieving the Sustainable Development Goals.

Excess Mortality due to Loiasis

Microfilaremic loiasis is associated with increased mortality in endemic populations. This was demonstrated in two studies in Cameroon and Congo (Chesnais et al. 2017; Hemilembolo et al. 2023). Both studies included data on the status of loiasis, the presence and extent of microfilaremia at baseline, and mortality data at follow-up after 15 and 17 years, respectively. The first study, conducted in Cameroon, described an increased mortality rate in hypermicrofilaremic individuals with more than 30,000 mf/ml and an age greater than 25 years at baseline. It was concluded that the extent of microfilaremia had a direct or indirect effect on increased mortality. Of note, the proportion of mortality attributable to the population associated with *L. loa* microfilaremia was high at 14.5%. This proportion is higher than the 5.2% reported for onchocerciasis (Chesnais et al. 2017). Interestingly, in the second study, excess mortality was found in all microfilaremic individuals, regardless of the extent of microfilaremia. This study from Congo showed that microfilaremic individuals died an average of 18.5 years earlier than amicrofilaremic individuals. The effect of microfilaremia on excess mortality was higher in individuals at younger ages at baseline but not at older ages. The crude mortality rate was 20.3 in Cameroon and 15.36 in Congo per 1000 person-years.

Hypotheses about the underlying mechanisms of excess mortality include the effects of severe organ damage associated with infection, chronic immune stimulation such as hypereosinophilia, or chronic immune modulation, which presents as immunosuppression and may carry a higher risk of acquiring other infections or diseases. Although future studies may help elucidate the underlying mechanisms of excess mortality, it is important to note that loiasis is associated with lower life expectancy. This underlines that loiasis is not a benign disease.

In summary, loiasis affects populations in tropical and subtropical areas and disproportionately affects rural and poorer populations in endemic regions. Loiasis causes significant morbidity and mortality and hinders the achievement of the SDGs.

Future estimates are likely to yield much higher DALYs, not only because they should take into account both morbidity and mortality but also include serious organ manifestations. In addition, the sociocultural impact of the disease should also be considered. Such data will allow for better estimates that reflect the true burden of disease and provide a much more realistic picture.

3 Local Aspects of Loiasis Disease

The endemicity of loiasis is geographically limited but includes regions where the disease is hyperendemic and more than 60% of the adult population have had an "eyeworm" episode at least once in their lifetime. To date, no formal analyses of disease perceptions and the sociocultural impact of loiasis have been published. However, such studies are of paramount importance in assessing the impact of a disease on affected populations and adjusting future control or treatment programs accordingly. Nevertheless, certain aspects can be discussed on the basis of the current state of knowledge.

Especially in areas of high endemicity, the disease is well known, as evidenced by specific names for the parasite in different languages and the availability of specific traditional treatments (Mischlinger et al. 2018; Takougang et al. 2007; Veletzky et al. 2021). Treatments for "eyeworm" include various herbs that are home-processed or prepared by traditional healers, liquids derived from plants or juices from foods such as chili peppers, onions, or garlic, and experimental treatments with urine, menthol, or gasoline (see Fig. 8). While these treatments have varying degrees of success, some result in permanent visual impairment or loss of sight (Veletzky et al. 2021). In addition, attempted and more or less successful removal of *Loa loa* worms with wooden sticks or needles is repeatedly reported in the literature (see Fig. 9). All these efforts lead to the same interpretation: The "eyeworm" episode is disturbing and is associated with disability and shame; in short, the affected patients

Fig. 8 Leaves used in Gabon for the preparation of traditional remedies against “eyeworm” (the subconjunctival migration of the *Loa loa* adult worm). The leaves were shown by a traditional healer who boiled the leaves for hours followed by various preparation steps. According to her, there are several herbal liquids that expel the worm, but her recipe would make the worm disappear. In addition to herbal tinctures, she also offered removal of the worm with a needle if the patient desired. Both methods were frequently used. Other treatments she offered included remedies for back pain, hemorrhoids, or headaches. Photo: J. Hergeth

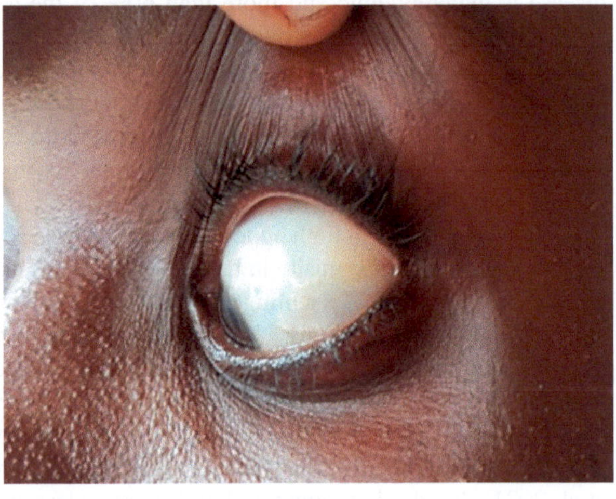

Fig. 9 This patient reported that her grandmother had unsuccessfully attempted to remove the worm from her eye years ago, leaving her with a residue of the dead worm in her eye that caused constant discomfort. Photo adapted from Veletzky et al. (2020)

want to get rid of it, indicating the importance of the parasite within the endemic population.

4 Conclusion

Loiasis is increasingly acknowledged as a disease of individual and public health relevance. Recent studies assessing disease burden and mortality have shown what has long been suggested by public health professionals, researchers, and clinicians working with the disease: Loiasis causes a range of debilitating symptoms and shortens life span. Next to the well-known and typical symptoms of "eyeworm" (the visible subconjunctival migration of the *Loa loa* adult worm) and Calabar swelling, many unspecific symptoms and a range of disease manifestations have been associated with the disease. Hypermicrofilariaemia with microfilaria loads of more than 20,000 mf/ml is a specificity of loiasis and is known to cause treatment-related but also spontaneous serious adverse events. *L. loa* microfilaremia is thought to be associated with organ damage, excess mortality, and development of atypical disease manifestations. The disease is well known in endemic regions and often treated with traditional remedies. While many open questions remain, it is thus evident that the disease is of relevance and should be acknowledged as a neglected tropical disease.

Acknowledgement We thank JP Akue, M Boussinesq, and Amy Klion for their input and revision of the chapter.

References

Akue JP, Hommel M, Devaney E (1997) High levels of parasite-specific IgG1 correlate with the amicrofilaremic state in Loa loa infection. J Infect Dis 175:158–163. https://doi.org/10.1093/infdis/175.1.158

Akue JP, Nkoghe D, Padilla C, Moussavou G, Moukana H, Mbou RA, Ollomo B, Leroy EM (2011) Epidemiology of concomitant infection due to Loa loa and Mansonella perstans in Gabon. PLoS Negl Trop Dis 5:e1329. https://doi.org/10.1371/journal.pntd.0001329

Andy JJ, Bishara FF, Soyinka OO, Odesanmi WO (1981) Loasis as a possible trigger of African endomyocardial fibrosis: a case report from Nigeria. Acta Trop 38:179–186

Andy JJ, Ogunowo PO, Akpan NA, Odigwe CO, Ekanem IA, Esin RA (1998) Helminth associated hypereosinophilia and tropical endomyocardial fibrosis (EMF) in Nigeria. Acta Trop 69:127–140

Beaver PC (1989) Intraocular filariasis: a brief review. Am J Trop Med Hyg 40:40–45. https://doi.org/10.4269/ajtmh.1989.40.40

Berenguer A, Plancha E, Gil JM (2003) Right ventricular endomyocardial fibrosis and microfilarial infection. Int J Cardiol 87:287–289. https://doi.org/10.1016/S0167-5273(02)00309-1

Bhalla D, Dumas M, Preux P-M (2013) Neurological manifestations of filarial infections. Handb Clin Neurol 114:235–242. https://doi.org/10.1016/B978-0-444-53490-3.00018-2

Bogaert LV, Dubois A, Janssens PG, Radermecker J, Tverdy G, Wanson M (1955) ENCEPHALITIS IN LOA-LOA FILARIASIS. J Neurol Neurosurg Psychiatry 18:103–119. https://doi.org/10.1136/jnnp.18.2.103

Bouchaud O, Matheron S, Loarec A, Dupouy Camet J, Bourée P, Godineau N, Poilane I, Cailhol J, Caumes E (2020) Imported loiasis in France: a retrospective analysis of 167 cases with comparison between sub-Saharan and non sub-Saharan African patients. BMC Infect Dis 20: 63. https://doi.org/10.1186/s12879-019-4740-6

Boussinesq M, Gardon J (1997) Prevalences of Loa loa microfilaraemia throughout the area endemic for the infection. Ann Trop Med Parasitol 91:573–589. https://doi.org/10.1080/00034989760671

Boussinesq M, Gardon J, Gardon-Wendel N, Chippaux J-P (2003) Clinical picture, epidemiology and outcome of Loa-associated serious adverse events related to mass ivermectin treatment of onchocerciasis in Cameroon. Filaria J 2(Suppl 1):S4. https://doi.org/10.1186/1475-2883-2-S1-S4

Boussinesq M, Gardon J, Gardon-Wendel N, Kamgno J, Ngoumou P, Chippaux JP (1998) Three probable cases of Loa loa encephalopathy following ivermectin treatment for onchocerciasis. Am J Trop Med Hyg 58:461–469

Bouyou Akotet MK, Owono-Medang M, Mawili-Mboumba DP, Moussavou-Boussougou MN, Nzenze Afène S, Kendjo E, Kombila M (2016) The relationship between microfilaraemic and amicrofilaraemic loiasis involving co-infection with Mansonella perstans and clinical symptoms in an exposed population from Gabon. J Helminthol 90:469–475. https://doi.org/10.1017/S0022149X15000607

Bouyou-Akotet MK, Moussavou Boussougou MN, Ovono-Abessolo F, Owono-Medang M, Kombila M (2014) Influence of Mansonella perstans microfilaraemia on total IgE levels in Gabonese patients co-infected with Loa loa. Acta Trop 131:11–15. https://doi.org/10.1016/j.actatropica.2013.11.012

Buell KG, Whittaker C, Chesnais CB, Jewell PD, Pion SDS, Walker M, Basáñez M-G, Boussinesq M (2019) Atypical clinical manifestations of loiasis and their relevance for endemic populations. Open Forum Infect Dis 6. https://doi.org/10.1093/ofid/ofz417

Burchard GD, Reimold-Jehle U, Bürkle V, Kretschmer H, Vierbuchen M, Racz P, Lo Y (1996) Splenectomy for suspected malignant lymphoma in two patients with loiasis. Clin Infect Dis 23: 979–982. https://doi.org/10.1093/clinids/23.5.979

Cambanis A (2010) Pulmonary loiasis and HIV coinfection in rural Cameroon. PLoS Negl Trop Dis 4:e572. https://doi.org/10.1371/journal.pntd.0000572

Carme B, Kaya-Gandziami G, Pintart D (1984) Localization of the filaria Loa loa in the anterior chamber of the eye. Apropos of a case. Acta Trop 41:265–269

Carme B, Mamboueni JP, Copin N, Noireau F (1989) Clinical and biological study of Loa loa filariasis in Congolese. Am J Trop Med Hyg 41:331–337

Carme B, Paraiso D, Gombe-Mbalawa C (1990) Calcifications of the breast probably due to Loa loa. Am J Trop Med Hyg 42:65–66. https://doi.org/10.4269/ajtmh.1990.42.65

Chang LW, Reller ME, Bishop JA, Talaat K, Nutman TB, Auwaerter PG (2008) A 41-year-old woman from Cameroon with infertility. Clin Infect Dis 47:141. https://doi.org/10.1086/588787

Chesnais CB, Takougang I, Paguélé M, Pion SD, Boussinesq M (2017) Excess mortality associated with loiasis: a retrospective population-based cohort study. Lancet Infect Dis 17:108–116. https://doi.org/10.1016/S1473-3099(16)30405-4

Churchill DR, Morris C, Fakoya A, Wright SG, Davidson RN (1996) Clinical and laboratory features of patients with loiasis (Loa loa filariasis) in the U.K. J Infect 33:103–109. https://doi.org/10.1016/s0163-4453(96)93005-4

da Silva MF, Saito AY, Peres VJ, Oliveira AC, Katzin AM (2015) In vitro antimalarial activity of different inhibitors of the plasmodial isoprenoid synthesis pathway. Antimicrob Agents Chemother 59:5084–5087. https://doi.org/10.1128/AAC.04161-14

Djikeussi ED, Akue JP (2014) Age-dependent prevalence of Loa loa amicrofilaremia and microfilaremia status as defined by two markers: microfilaria and specific IgG4. AJB 13:593–597. https://doi.org/10.5897/AJB09.645

Dupont A, Zue-N'dong J, Pinder M (1988) Common occurrence of amicrofilaraemic Loa loa filariasis within the endemic region. Trans R Soc Trop Med Hyg 82:730

Eyebe S, Sabbagh A, Pion SD, Nana-Djeunga HC, Kamgno J, Boussinesq M, Chesnais CB (2018) Familial aggregation and heritability of Loa loa microfilaremia. Clin Infect Dis 66:751–757. https://doi.org/10.1093/cid/cix877

Gantois N, Rapp C, Gautret P, Ficko C, Savini H, Larreché S, Saidi R, Crevon L, Simon F (2013) Imported loiasis in France: a retrospective analysis of 47 cases. Travel Med Infect Dis, Special Issue: Skin 11:366–373. https://doi.org/10.1016/j.tmaid.2013.08.005

Gardon J, Gardon-Wendel N, Demanga-Ngangue, Kamgno J, Chippaux JP, Boussinesq M (1997) Serious reactions after mass treatment of onchocerciasis with ivermectin in an area endemic for Loa loa infection. Lancet 350:18–22. https://doi.org/10.1016/S0140-6736(96)11094-1

Gobbi F, Boussinesq M, Mascarello M, Angheben A, Gobbo M, Rossanese A, Corachán M, Bisoffi Z (2011) Loiasis with peripheral nerve involvement and spleen lesions. Am J Trop Med Hyg 84: 733–737. https://doi.org/10.4269/ajtmh.2011.10-0458

Hall CL, Stephens L, Peat D, Chiodini PL (2001) Nephrotic syndrome due to loiasis following a tropical adventure holiday: a case report and review of the literature. Clin Nephrol 56:247–250

Hassan S, Isyaku M, Yayo A, Sarkin Fada F, Ihesiulor GU, Iliyasu G (2016) Adult Loa loa filarial worm in the anterior chamber of the eye: a first report from Savanna Belt of Northern Nigeria. PLoS Negl Trop Dis 10:e0004436. https://doi.org/10.1371/journal.pntd.0004436

Hautekeete ML, Pailoux G, Marcellin P, Girard PM, Degott C, Benhamou JP (1989) Presence of Loa loa microfilariae in ascitic fluid. J Infect Dis 160:559–560. https://doi.org/10.1093/infdis/160.3.559a

Hawking F, Moore P, Gammage K, Worms MJ (1967) Periodicity of microfilariae. XII. The effect of variations in host body temperature on the cycle of Loa loa, Monnigofilaria setariosa, Dirofilaria immitis and other filariae. Trans R Soc Trop Med Hyg 61:674–683. https://doi.org/10.1016/0035-9203(67)90132-0

Hemilembolo MC, Niama AC, Campillo JT, Pion SD, Missamou F, Whittaker C, Kankou J-M, Ndziessi G, Bileckot RR, Boussinesq M, Chesnais CB (2023) Excess mortality associated with loiasis: confirmation by a new retrospective cohort study conducted in the Republic of Congo. Open Forum Infect Dis 10:ofad103. https://doi.org/10.1093/ofid/ofad103

Herrick JA, Makiya MA, Holland-Thomas N, Klion AD, Nutman TB (2020) Infection-associated immune perturbations resolve 1 year following treatment for Loa loa. Clin Infect Dis 72:789–796. https://doi.org/10.1093/cid/ciaa137

Herrick JA, Metenou S, Makiya MA, Taylar-Williams CA, Law MA, Klion AD, Nutman TB (2015) Eosinophil-associated processes underlie differences in clinical presentation of loiasis between temporary residents and those indigenous to Loa-endemic areas. Clin Infect Dis 60:55–63. https://doi.org/10.1093/cid/ciu723

Hulin C, Rabaud C, May T, Neimann L, Kures L, Canton P (1994) Pulmonary involvement with a favorable course during Loa loa filariasis. Bull Soc Pathol Exot 87:248–250

Jaffres R, Simitzis Le Flohic AM, Chastel C (1983) Loa loa filarial arthritis with microfiliara in the articular fluid. Rev Rhum Mal Osteoartic 50:145–147

Katner H, Beyt BE, Krotoski WA (1984) Loiasis and renal failure. South Med J 77:907–908. https://doi.org/10.1097/00007611-198407000-00027

Klion AD, Eisenstein EM, Smirniotopoulos TT, Neumann MP, Nutman TB (1992) Pulmonary involvement in loiasis. Am Rev Respir Dis 145:961–963. https://doi.org/10.1164/ajrccm/145.4_Pt_1.961

Klion AD, Massougbodji A, Sadeler BC, Ottesen EA, Nutman TB (1991) Loiasis in endemic and nonendemic populations: immunologically mediated differences in clinical presentation. J Infect Dis 163:1318–1325

León D, Martín M, Corros C, Santamarta E, Costilla S, Lambert JL (2012) Usefulness of cardiac MRI in the early diagnosis of endomyocardial fibrosis. Rev Port Cardiol 31:401–402. https://doi.org/10.1016/j.repce.2012.04.013

Lukiana T, Mandina M, Situakibanza NH, Mbula MM, Lepira BF, Odio WT, Kamgno J, Boussinesq M (2006) A possible case of spontaneous Loa loa encephalopathy associated with a glomerulopathy. Filaria J 5:6. https://doi.org/10.1186/1475-2883-5-6

Metzger WG, Mordmüller B (2014) Loa loa-does it deserve to be neglected? Lancet Infect Dis 14: 353–357. https://doi.org/10.1016/S1473-3099(13)70263-9

Mischlinger J, Veletzky L, Tazemda-Kuitsouc GB, Pitzinger P, Matsegui PB, Gmeiner M, Lagler H, Gebru T, Held J, Mordmüller B, Ramharter M (2018) Behavioural and clinical predictors for Loiasis. J Glob Health 8:010413. https://doi.org/10.7189/jogh.08.010413

Mitre E, Klion AD (2021) Eosinophils and helminth infection: protective or pathogenic? Semin Immunopathol 43:363–381. https://doi.org/10.1007/s00281-021-00870-z

Mombo-Ngoma G, Mackanga JR, Basra A, Capan M, Manego RZ, Adegnika AA, Lötsch F, Yazdanbakhsh M, González R, Menendez C, Mabika B, Matsiegui PB, Kremsner PG, Ramharter M (2015) Loa loa infection in pregnant women, Gabon. Emerging Infect Dis 21: 899–8901. https://doi.org/10.3201/eid2105.141471

Mongin A (1770) Sur un ver trouvé sous la conjonctive à Maribarou, isle Saint-Dominique. J Med Chir Pharm Paris 32:338–339

Murray CJ, Acharya AK (1997) Understanding DALYs (disability-adjusted life years). J Health Econ 16:703–730

Nana-Djeunga HC, Fossuo-Thotchum F, Pion SD, Chesnais CB, Kubofcik J, Mackenzie CD, Klion AD, Boussinesq M, Nutman TB, Kamgno J (2019) Loa loa microfilariae in skin snips: consequences for onchocerciasis monitoring and evaluation in L. loa–Endemic Areas. Clin Infect Dis 69:1628. https://doi.org/10.1093/cid/ciz172

Neafie RC, Negesse Y, Connor DH, Lanoie LO (1985) Loiasis: "Calabar" swellings and involvement of deep organs*. Am J Trop Med Hyg 34:537–546. https://doi.org/10.4269/ajtmh.1985.34.537

Ngu JL, Chatelanat F, Leke R, Ndumbe P, Youmbissi J (1985) Nephropathy in Cameroon: evidence for filarial derived immune-complex pathogenesis in some cases. Clin Nephrol 24:128–134

Ngu JL, Youmbissi TJ (1987) Special features, pathogenesis and aetiology of glomerular diseases in the tropics. Clin Sci 72:519–524. https://doi.org/10.1042/cs0720519

Noireau F, Apembet JD, Nzoulani A, Carme B (1990) Clinical manifestations of loiasis in an endemic area in the Congo. Trop Med Parasitol 41:37–39

Ntsame Ondo N (2015) Direction Générale de la Statistique (2015) : Résultats globaux du Recensement Général de la Population et des Logements de 2013 du Gabon (RGPL-2013). Libreville

Nutman TB, Miller KD, Mulligan M, Ottesen EA (1986) Loa loa infection in temporary residents of endemic regions: recognition of a hyperresponsive syndrome with characteristic clinical manifestations. J Infect Dis 154:10–18

Pakasa NM, Nseka NM, Nyimi LM (1997) Secondary collapsing glomerulopathy associated with Loa loa filariasis. Am J Kidney Dis 30:836–839. https://doi.org/10.1016/s0272-6386(97)90090-1

Penel R (1905) Les filaires du sang de l'homme. Paris:127–140

Pillay VKG (1973) Glomerulopathy associated with filarial loiasis. JAMA 225:179. https://doi.org/10.1001/jama.1973.03220290057028

Pinder M (1988) Loa loa – a neglected filaria. Parasitol Today (Regul Ed) 4:279–284

Pion DSS, Gardon J, Kamgno J, Gardon-Wendel N, Chippaux JP, Boussinesq M (2004) Structure of the microfilarial reservoir of Loa loa in the human host and its implications for monitoring the programmes of Community-Directed Treatment with Ivermectin carried out in Africa. Parasitology 129:613–626. https://doi.org/10.1017/s0031182004005694

Pion SDS, Demanou M, Oudin B, Boussinesq M (2005) Loiasis: the individual factors associated with the presence of microfilaraemia. Ann Trop Med Parasitol 99:491–500. https://doi.org/10.1179/136485905X51300

Puy V, Scheffler F, Tinez C, Belhadri-Mansouri N, Huguenin A, Capelle S, Agnamey P, Devaux A, Frydman N, Copin H, Benkhalifa M (2017) Detection of Loa loa microfilariae in follicular fluid during assisted reproductive technology: case report and review of the literature. Open Forum Infect Dis 4:ofx208. https://doi.org/10.1093/ofid/ofx208

Rosenberg HF, Dyer KD, Foster PS (2013) Eosinophils: changing perspectives in health and disease. Nat Rev Immunol 13:9–22. https://doi.org/10.1038/nri3341

Roussel F, Roussel C, Brasseur P, Gourmelen O, Le Loet X (1989) Aseptic knee effusion with Loa loa microfilariae in the articular fluid. Acta Cytol 33:281–283

Tai PC, Ackerman SJ, Spry CJ, Dunnette S, Olsen EG, Gleich GJ (1987) Deposits of eosinophil granule proteins in cardiac tissues of patients with eosinophilic endomyocardial disease. Lancet 1:643–647. https://doi.org/10.1016/s0140-6736(87)90412-0

Takougang I, Meli J, Lamlenn S, Tatah PN, Ntep M (2007) Loiasis—a neglected and underestimated affliction: endemicity, morbidity and perceptions in eastern Cameroon. Ann Trop Med Parasitol 101:151–160. https://doi.org/10.1179/136485907X154511

Takougang I, Meremikwu M, Wandji S, Yenshu EV, Aripko B, Lamlenn SB, Eka BL, Enyong P, Meli J, Kale O, Remme JH (2002) Rapid assessment method for prevalence and intensity of Loa loa infection. Bull World Health Organ 80:852–858

Tamarozzi F, Buonfrate D, Ricaboni D, Ursini T, Foti G, Gobbi F (2022) Spleen nodules in Loa loa infection: re-emerging knowledge and future perspectives. Lancet Infect Dis. https://doi.org/10.1016/S1473-3099(21)00632-0

Védy J et al (1975) MANIFESTATIONS OCULAIRES ATYPIQUES DES FILARIOSES A LOA LOA

Veletzky L, Eberhardt KA, Hergeth J, Stelzl DR, Manego RZ, Mombo-Ngoma G, Kreuzmair R, Burger G, Adegnika AA, Agnandji ST, Matsiegui PB, Boussinesq M, Mordmüller B, Ramharter M (2022) Distinct loiasis infection states and associated clinical and hematological manifestations in patients from Gabon. PLoS Negl Trop Dis 16:e0010793. https://doi.org/10.1371/journal.pntd.0010793

Veletzky L, Hergeth J, Stelzl DR, Mischlinger J, Manego RZ, Mombo-Ngoma G, McCall MBB, Adegnika AA, Agnandji ST, Metzger WG, Matsiegui PB, Lagler H, Mordmüller B, Budke C, Ramharter M (2020) Burden of disease in Gabon caused by loiasis: a cross-sectional survey. Lancet Infect Dis. https://doi.org/10.1016/S1473-3099(20)30256-5

Veletzky L, Mombo-Ngoma G, Ramharter M (2021) The supposedly innocuous eye worm: or, what could make you put chili in your eye? Am J Trop Med Hyg 105:1448–1449. https://doi.org/10.4269/ajtmh.21-0319

Wanji S, Eyong E-EJ, Tendongfor N, Ngwa CJ, Esuka EN, Kengne-Ouafo AJ, Datchoua-Poutcheu FR, Enyong P, Agnew D, Eversole RR, Hopkins A, Mackenzie CD (2017) Ivermectin treatment of Loa loa hyper-microfilaraemic baboons (Papio anubis): assessment of microfilarial load reduction, haematological and biochemical parameters and histopathological changes following treatment. PLoS Negl Trop Dis 11:e0005576. https://doi.org/10.1371/journal.pntd.0005576

Whitaker D, Reed WD, Shilkin KB (1980) A case of filariasis diagnosed on gastric cytology. Pathology 12:483–486. https://doi.org/10.3109/00313028009077113

Williams I (1954) Calcification in loiasis. J Facul Radiologists 6:142–144. https://doi.org/10.1016/S0368-2242(54)80067-4

Wisanto A, Laureys M, Camus M, Devroey P, Verheyen G, van Steirteghem AC (1993) CASE REPORT: Loa loa microfilariae aspirated during oocyte retrieval. Hum Reprod 8:2096–2097. https://doi.org/10.1093/oxfordjournals.humrep.a137988

World Urbanization Prospects – Population Division – United Nations [WWW Document] (2018). https://population.un.org/wup/Country-Profiles/. Accessed 11 Jul 2019

Clinical Aspects: Treatment of Simple and Complicated Forms of Loiasis

Amy Klion

Abstract

The filarial parasite, *Loa loa,* causes chronic infection in an estimated 10 million people living in Central and West Africa and in long-term visitors to endemic areas. Although generally perceived as a relatively benign infection, the symptoms of loiasis can be disabling and infection has been associated with excess mortality and rare but serious complications of infection-associated eosinophilia. Treatment of loiasis is complex, due to potentially severe complications associated with microfilarial killing, especially in patients with high-microfilarial loads, and incomplete efficacy of macrofilaricidal therapies. In the following chapter, currently available therapies for loiasis and their use in the treatment of loiasis are reviewed. Novel therapies and adjunct measures are also discussed. Finally, a simple stepwise approach to treatment is suggested.

Keywords

Loa loa · Loiasis · Treatment · Diethylcarbamazine · Ivermectin · Albendazole

1 General Principles

Spread by tabanid flies of the *Chrysops* genus, the filarial parasite, *Loa loa,* causes chronic infection in an estimated 10 million people living in Central and West Africa and in long-term visitors to endemic areas. Whether to treat an individual patient and with what agent depends on many factors, including the severity of clinical

A. Klion (✉)
Human Eosinophil Section, Laboratory of Parasitic Diseases, National Institutes of Allergy and Infectious Diseases, National Institutes of Health, Bethesda, MD, USA
e-mail: aklion@niaid.nih.gov

J. P. Akue (ed.), *Loa loa: Latest Advances in Loiasis Research*,
https://doi.org/10.1007/978-3-031-49450-5_5

manifestations, the presence and level of circulating blood microfilariae, concomitant infections, the likelihood of reinfection, and access to therapy.

1.1 Clinical Manifestations

The clinical manifestations of *L. loa* infection include asymptomatic microfilaremia, transient migratory angioedema (Calabar swellings), subconjunctival migration of an adult worm (eyeworm), and nonspecific symptoms such as fatigue, urticaria and myalgia/arthralgia. These symptoms can be debilitating, as evidenced by a cross-sectional study in rural Gabon which estimated that morbidity due to loiasis accounted for 412.9 disability-adjusted life years (DALYs) per 100,000 people (Veletzky et al. 2020). More serious complications of untreated loiasis have been reported and include fatal encephalitis (Kivits 1952; Van Bogaert et al. 1955; Arrey-Agbor et al. 2018), renal failure (Ngu et al. 1985; Pakasa et al. 1997), and blindness (Buell et al. 2019). Moreover, hypereosinophilia ($>1.5 \times 10^9$/L) is common in *L. loa* infection (Herrick et al. 2015; Bottieau et al. 2022; Veletzky et al. 2022) and can be associated with rare but serious sequelae, including potentially fatal endomyocardial fibrosis (Nutman et al. 1986; Andy et al. 1998; Herrick et al. 2015). Whereas numerous studies have demonstrated population differences in clinical presentation between residents of endemic areas and temporary visitors to the same areas (Nutman et al. 1986; Churchill et al. 1996; Gantois et al. 2013; Herrick et al. 2015; Saito et al. 2015; Bouchaud et al. 2020), individual patients may fall anywhere within the spectrum of clinical presentations irrespective of their demographics.

1.2 Treatment Indications

Treatment has traditionally been recommended for patients with significant symptoms and those with high-grade eosinophilia (due to the risk of eosinophil-driven complications). Although the blood-borne microfilariae themselves do not typically cause symptoms in untreated patients, serious complications, including encephalopathy (Kivits 1952), have been reported in the absence of eosinophilia in patients with high-microfilarial levels. More importantly, recent data suggest that the presence of *L. loa* microfilaremia is associated with an increased mortality risk (Chesnais et al. 2017; Hemilembolo et al. 2023). Thus, treatment is potentially beneficial for all infected individuals and must be weighed against the potential risks of posttreatment reactions.

1.3 Complications of Treatment

Complications of treatment for loiasis fall into three major categories: (1) drug-specific toxicity (discussed in the treatment-specific sections below), (2) side effects related to effects of the drug on concomitant infections (discussed in the treatment-specific sections below), and (3) the consequences of drug activity against the different stages of the *L. loa* parasite. Whereas killing of the adult worm can provoke local reactions (e.g., inflammatory nodules, angioedema), these reactions are self-limited and rarely severe. In contrast, rapid killing of microfilariae can lead to significant complications especially in patients with high levels of circulating parasites. In a recent study that used statistical modeling to assess individual risk following ivermectin administration, it was estimated that 1% of people with 20,000 microfilariae/mL of blood and 28% of those with 100,000 microfilariae/mL of blood will develop a serious adverse event (Chesnais et al. 2020). Although neurologic events are the most frequently reported serious complication of microfilaricidal therapy, other rare but clinically significant sequelae include renal failure (Cruel et al. 1997) and pulmonary complications, such as pleural effusion (Carme et al. 1982; Klion et al. 1992).

After multiple case descriptions implicating diethylcarbamazine (DEC) as a cause of encephalitis in loiasis (Van Bogaert et al. 1955; Stanley and Kell 1982), Carme et al. (1991a) reported a series of 11 cases, of which five were fatal, clearly documenting the relationship between DEC administration and neurologic complications in patients with *L. loa* microfilaremia. Reports of similar events in the setting of mass drug treatment programs using ivermectin to eliminate transmission of onchocerciasis and lymphatic filariasis in *L. loa*—endemic areas in Africa (Gardon et al. 1997a) and the identification of *Loa loa* microfilariae in the cerebrospinal fluid of four of ten patients with *L. loa* blood microfilarial levels >10,000/mL following (but not before) ivermectin therapy (Ducorps et al. 1995) resulted in suspension of ivermectin-based mass drug administration programs in several regions (Makenga Bof et al. 2019). These events sparked renewed interest in understanding the pathophysiology of posttreatment encephalopathy in loiasis.

Histopathological studies of the brain in patients with loiasis and posttreatment neurologic complications have been few but have typically shown meningitis and vascular lesions with fibrin thrombi, degenerating microfilariae and inflammatory cells (including in one case eosinophils) (Van Bogaert et al. 1955; Cauchie et al. 1965; Negesse et al. 1985). These findings were confirmed in a study using a splenectomized baboon model of hypermicrofilaremic loiasis (Wanji et al. 2015). Twelve baboons with high-microfilarial loads (19,800–124,700/mL of blood) were divided into four groups: no treatment, single-dose ivermectin alone, single-dose ivermectin followed 3 days later by aspirin (500 mg twice daily for 2 days), and single-dose ivermectin followed 3 days later by prednisone (20 mg twice daily for 2 days) (Wanji et al. 2017). Ivermectin dramatically reduced the microfilarial counts in all treated animals (by 62–98%). Moreover, microfilarial reduction was accompanied by neurologic and dermatologic symptoms and a rise in blood eosinophil counts. At necropsy on day 10, pathology revealed degenerating microfilariae and eosinophils in brain tissue with blockage of vascular capillaries. These data are consistent with the hypothesis that drug-induced paralysis or killing of the microfilariae leads to their entrapment in blood capillaries and susceptibility to immune-mediated destruction. The association between palpebral conjunctival and retinal hemorrhages following ivermectin therapy and blood microfilarial levels provides additional support for an underlying vascular process (Fobi et al. 2000). Whereas data demonstrating microfilariae in the cerebrospinal fluid posttreatment are convincing, the role that they play in the pathophysiology of posttreatment encephalopathy remains unclear.

Several studies have shown that the peripheral blood absolute eosinophil count (AEC) increases approximately 24 h following administration of DEC or ivermectin to patients with loiasis (Martin-Prevel et al. 1993; Ducorps et al. 1995; Herrick et al. 2017; Legrand et al. 2021). This posttreatment increase in AEC is preceded by a dramatic increase in serum levels of the cytokine, IL-5, and accompanied by evidence of eosinophil activation, including release of eosinophil granule proteins (Herrick et al. 2017; Legrand et al. 2021). Eosinophil granule proteins are cationic proteins with substantial cytotoxic activity that have been implicated in the pathogenesis of hypereosinophilic syndromes and in the Mazzotti reaction following DEC treatment of onchocerciasis (Ackerman et al. 1990). These data together with the histopathologic findings described above suggest that eosinophils may play an important role in the pathogenesis of posttreatment reactions in loiasis. Although a small placebo-controlled randomized trial using the monoclonal anti-IL-5 antibody, reslizumab, to block eosinophilia following DEC treatment in patients with loiasis failed to demonstrate a reduction in posttreatment symptoms, the size of the study population was small and reslizumab blunted but failed to completely prevent posttreatment eosinophilia and eosinophil activation (Legrand et al. 2021).

Whereas more definitive studies are clearly needed to better understand the pathophysiologic mechanisms underlying posttreatment reactions in loiasis, recent studies using the LoaScope, a cellphone-based microscope that allows accurate point-of-care assessment of blood microfilarial levels, have confirmed the association between serious adverse events and high-microfilarial density. In a large community-based study (Test and not Treat) in a district coendemic for onchocerciasis and loiasis in Cameroon, individuals with microfilarial loads >20,000/mL were excluded prior to ivermectin distribution to the remaining eligible population (n = 15,522) (Kamgno et al. 2017). Although mild to moderate side effects were reported in 6% of treated individuals and were more common in the setting of *Loa loa* microfilaremia and/or positive Ov16 serology for onchocerciasis, no serious adverse events were reported. Of note, 23 cases of encephalopathy, and three deaths, occurred during ivermectin distribution to 6000 residents of this district in 1999 (Haselow et al. 2003). The LoaScope was subsequently used to demonstrate that the lack of serious reactions during ivermectin mass drug administration in 110 villages in southern Nigeria was due to unexpectedly low-microfilarial levels in this population (Emukah et al. 2018).

Efforts to identify risk factors for serious posttreatment reactions other than the degree of *L. loa* microfilaremia have found no evidence that risk is affected by parasite strain (Higazi et al. 2004; Kamgno et al. 2009), mdr-1 status (Bourguinat et al. 2010), or coinfection with *Mansonella perstans* or *Plasmodium* (Gardon et al. 1997a; Fokom-Domgue et al. 2014; Chesnais et al. 2020). A recent study found that males were three times more likely to develop a serious adverse event following ivermectin therapy (Chesnais et al. 2020), although the reasons for this are unknown.

1.4 Other Factors That Affect Treatment Decisions

The three drugs currently available for treatment of loiasis (DEC, ivermectin, and albendazole), have different mechanisms of action, effects on the different life cycle stages of the parasite, and safety concerns (Table 1). Whereas clinical manifestations and blood microfilarial levels are clearly important in the decision to treat and choice of therapeutic agent, other factors to consider include the likelihood of reinfection (i.e., the relative benefit of curative therapy vs. suppressive therapy), access to the

Table 1 Drugs currently used for the treatment of loiasis

	DEC	Ivermectin	Albendazole
Manufacturer	Multiple	Multiple	Multiple
Drug class	Piperazine	Avermectin	Benzimidazole
Formula	$C_{10}H_{21}N_3O$	$C_{48}H_{74}O_{14}$	$C_{12}H_{15}N_3O_2S$
Structure	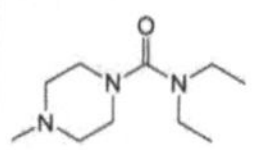	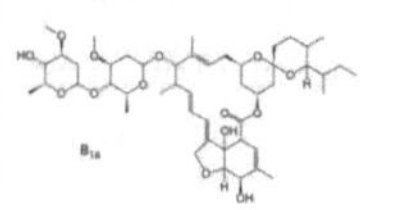	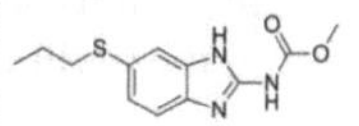
Mechanism of action	Incompletely understood but likely involves temporary paralysis through activation of TRP-2 and SLO-1 channels leading to trapping and destruction by host immune system; possible role for nitric oxide or other pathways	Binds to glutamate-gated chloride ion channels in invertebrate nerve and muscle cells leading to decreased protein and extracellular vesicle secretion	Binds to tubulin inhibiting the formation of microtubules resulting in impaired embryogenesis and parasite death
Recommended dose for treatment of loiasis	8 mg/kg orally in three divided doses for 21 days; dose escalation beginning with 50 mg test dose may reduce side effects; contraindicated in patients with high-microfilarial loads (>8000/mL)	150–200 mcg/kg as a single oral dose with or without concomitant macrofilaricidal therapy; contraindicated in patients with microfilarial loads >20,000/mL	200–400 mg orally twice daily for 3 weeks
Microfilaricidal activity	Yes	Yes	No
Macrofilaricidal activity	Yes	No	Yes
Utility in prophylaxis	Yes	No	No
Safety concerns	Posttreatment effects related to effect on microfilariae	Posttreatment effects related to effect on microfilariae	Liver and hematologic toxicity

different medications, cost of therapy and/or monitoring, and convenience (e.g., duration of therapy, number of pills or tablets).

2 Antifilarial Drugs

2.1 Diethylcarbamazine

Diethylcarbamazine (1-diethylcarbamyl-4-methylpiperazine hydrochloride) is a piperazine derivative developed by American Cyanamid Company that was first found to have antifilarial activity in animal models in the mid-1940s (Hewitt and White 1947). After a clinical trial in 27 patients with lymphatic filariasis demonstrated rapid and sustained microfilarial clearance and some evidence of activity against adult worms without significant toxicity (Santiago-Stevenson et al. 1947), the safety and efficacy of DEC were explored in patients with loiasis. Although the earliest reports of DEC treatment of loiasis describe mild to moderate exacerbations of clinical symptoms without other evidence of toxicity (Shookhoff and Dwork 1949; Murgatroyd and Woodruff 1949), the patients treated in these studies were expatriates who acquired their infection during long-term stays in endemic areas and had no or few detectable blood microfilariae. That said, the authors described clearance of microfilariae and resolution of clinical symptoms in all 22 patients prompting further studies of DEC for the treatment of loiasis.

DEC is one of the three antifilarial agents on the 2021 WHO Model List of Essential Medicines and is commercially available in many countries. Previously approved by the Food and Drug Administration in a different formulation, DEC is currently available for use in the United States for the treatment of loiasis in adults and children older than 18 months of age only from the Centers for Disease Control under an Investigational New Drug protocol. DEC is not recommended for use in pregnancy due to the paucity of outcomes data in patients inadvertently treated during pregnancy.

2.1.1 Mechanism of Action

The mechanism of action of DEC remains incompletely understood. The rapid *in vivo* clearance of microfilariae from the blood of DEC-treated mice infected with the filarial parasite, *Litosomoides carinii*, accumulation of disintegrating microfilariae surrounded by phagocytes in the liver of treated mice, and the lack of effect of DEC on isolated *Litosomoides* microfilariae in culture led to the early hypothesis that DEC alters the microfilariae in such a way that they become trapped in the liver and destroyed (Hawking et al. 1950). Destruction of *Loa loa* microfilariae in the liver following DEC therapy was subsequently demonstrated in a Nigerian man with loiasis who underwent sequential biopsies (Woodruff 1951). The very recent finding that low concentrations of DEC open transient receptor protein (TRP) channels in *Brugia malayi* adult worms and microfilariae leading to activation of calcium-dependent SLOw poke potassium channels (SLO-1) in somatic muscle cells and temporary paralysis supports this hypothesis (Verma et al. 2020). Moreover, it provides a potential explanation for the requirement for components of the host immune response in DEC-mediated killing of *Loa loa* and *Brugia malayi* microfilariae *in vitro* (Piessens and Beldekas 1979; Cesbron et al. 1987).

DEC has also been shown to have effects on arachidonic acid metabolism and prostaglandin production (Kanesa-thasan et al. 1991; Maizels and Denham 1992). Although neither *C. elegans* nor *Brugia malayi* worms appear to have homologs of cyclooxygenases or lipoxygenases (Verma et al. 2020), they do have CYP450 cytochrome oxidases, including epoxygenases and omega-hydroxylases, that act on arachidonic acid to produce polyunsaturated fatty acids that are, in turn, involved in the production of prostaglandins (Liu et al. 1990). Whereas arachidonic acid appears to reduce TRP-2 channel activation under normal conditions, inhibition of epoxygenases has been shown to divert arachidonic acid metabolism to the production of metabolites that activate TRP-2 channels (Verma et al. 2020). Of note, TRP channels have also been described on mammalian cells, including monocytes and macrophages where they have been shown to inhibit the generation of reactive oxygen intermediates (Santoni et al. 2018). Induction of nitric oxide (NO) has been proposed as a potential mechanism of the filaricidal action of DEC in animal models; however, data have been conflicting. Early experiments demonstrated a reduction in cyclooxygenase levels in peritoneal exudate cells from mice treated with DEC and a lack of effect of DEC on microfilarial levels in inducible nitric oxide synthetase (iNOS)-deficient mice (McGarry et al. 2005), but subsequent studies failed to show a direct effect of DEC on iNOS or NO production by murine or rat macrophages or endothelial *in vitro* or urine nitrate excretion by rats administered DEC *in vivo* (Rajan et al. 1998). Finally, a recent study of the effects of high-dose DEC on young adult *C. elegans* suggests a second mechanism of action distinct from effects on TRP-2 (Datta et al. 2022).

2.1.2 Efficacy in Loiasis

As mentioned above, the first studies of DEC in loiasis were conducted in patients with relatively low levels of microfilaremia and demonstrated microfilarial clearance and resolution of symptoms, suggesting an effect on adult worms (Shookhoff and Dwork 1949; Murgatroyd and Woodruff 1949). Further evidence for macrofilaricidal activity of DEC in loiasis includes data from a trial in five mandrills where DEC treatment led to a 0–66% reduction in adult worms (Duke 1990), the finding of adult worms in biopsies of posttreatment nodules from patients with loiasis treated with DEC (Murgatroyd and Woodruff 1949; Nutman et al. 1986), and the large body of clinical data supporting the efficacy of DEC treatment in promoting "cure" in infected patients (Table 2). Although reported cure rates following a single course of DEC range from 38% to 94% depending on the study, differences in patient populations, dosing, length of follow-up, and the definition of cure likely account for most of this variability (Table 2). Unfortunately, data from prospective trials and validated objective measures of cure are lacking. Repeat treatment courses are frequently, but not always, effective in achieving cure in patients who relapse (Klion et al. 1994; Churchill et al. 1996).

Table 2 Retrospective analyses of DEC treatment efficacy in loiasis

Reference	Population	Treatment[a]	Months of follow-up (range)	Cure rate[b]	Definition of cure
Herrick et al. (2021)	Travelers and immigrants ($n = 204$)	DEC 200 mg twice daily for 3 weeks	>12	55% with a single course (99% after 2–4 courses)	Resolution of symptoms, microfilaremia, and eosinophilia at last follow-up
Bottieau et al. (2022)	Travelers and immigrants without reexposure ($n = 102$)	DEC 200 mg twice daily for 3 weeks (±IVM)	NA (3–12)	70%	Resolution of symptoms, microfilaremia, and eosinophilia at 3–12 months
Bouchaud et al. (2020)	Travelers and immigrants ($n = 36$)	DEC 200–400 mg daily for 3 weeks ($n = 10$)	Mean 6 (1–34)	0%	Resolution of symptoms and reduction of microfilaremia at last follow-up
		DEC 200–400 mg daily for 3 weeks then IVM ($n = 26$)		83%	
Gobbi et al. (2018)	Travelers and immigrants ($n = 98$)	DEC ($n = 74$)	Median 4.5	A:69%; B:50%	Resolution of symptoms (A); clearance of microfilaremia; and resolution of eosinophilia (B)
		DEC+IVM ($n = 16$)		A:71%; B:44%	
		DEC+ALB ($n = 8$)		A:100%; B:38%	
Saito et al. (2015)	Travelers and immigrants ($n = 36$)	DEC 600 mg daily for 3 weeks ($n = 33$)	Median 3	63%	No need for retreatment
		IVM then DEC 600 mg daily for 3 weeks ($n = 3$)		100%	
Gantois et al. (2013)	Travelers and immigrants ($n = 11$)	DEC	Mean 4.7	70%	Resolution of symptoms
Churchill et al. (1996)	Hospitalized travelers and immigrants ($n = 95$)	DEC	Median 3 (0–135)	88%	No need for retreatment

(continued)

Table 2 (continued)

Reference	Population	Treatment[a]	Months of follow-up (range)	Cure rate[b]	Definition of cure
Klion et al. (1994)	Travelers without reexposure ($n = 32$)	DEC (8 mg/kg/day for 3 weeks	Median 54 (24–180)	38%	Resolution of symptoms
Murgatroyd and Woodruff (1949)	Travelers ($n = 17$)	DEC 2–6 mg/kg/day for 10–21 days	NA (1–14)	94%	Resolution of symptoms

[a]*DEC* diethylcarbamazine, *IVM* ivermectin, *ALB* albendazole. If no dosing is given, this was not specified

[b]When this information was provided, patients with unknown follow-up were removed from the denominator in calculating % cure

Separate from its macrofilaricidal activity (which ultimately leads to reduction of *Loa loa* microfilarial levels), DEC has a rapid microfilaricidal effect, with complete clearance of blood microfilariae observed at 24–48 h posttreatment in most, if not all, treated individuals (Murgatroyd and Woodruff 1949; Herrick et al. 2017; Legrand et al. 2021). Reappearance of blood microfilariae can occur weeks to months after seemingly successful DEC treatment if residual adult worms are present or reinfection occurs. Finally, DEC also has efficacy against infective larvae and can be used to prevent acquisition of loiasis in travelers to endemic areas (see Sect. 5—Prevention and Prophylaxis).

2.1.3 Safety Concerns

Whereas mild to moderate side effects of DEC treatment, including pruritus, transient hematuria, and musculoskeletal symptoms, are frequent in patients with loiasis (Churchill et al. 1996; Herrick et al. 2021), serious adverse events related to DEC therapy are rare in travelers who acquire loiasis, consistent with the low prevalence of travelers who present with high-blood microfilarial levels (Nutman et al. 1986). In contrast, administration of DEC to residents of endemic areas can provoke serious, sometimes fatal, complications in patients with high levels of microfilaremia (see "Complications of treatment" above). These adverse effects do not occur in uninfected individuals treated with DEC (Murgatroyd and Woodruff 1949; Bolla et al. 2002) and are temporally related to the clearance of microfilariae from the blood. Attempts to reduce the risk of serious complications of therapy through dose modification, administration of glucocorticoids or other immunosuppressive have been ineffective. In contrast, reduction of microfilaremia prior to treatment either through apheresis or pretreatment with albendazole or ivermectin, appears to reduce the incidence of posttreatment complications related to killing of *L. loa* microfilariae. Concomitant infection with *Onchocerca volvulus* (onchocerciasis) poses an additional risk to DEC treatment of patients with loiasis due to the induction of severe, sometimes fatal, systemic inflammatory responses (Mazzotti reactions) in patients with skin and eye microfilariae (Hawking and Laurie 1949).

2.2 Ivermectin

Ivermectin is a semisynthetic derivative of avermectin, a fermentation product of *Streptomyces avermitilis*. First developed as a veterinary anthelminthic in 1975 by Merck, it is one of the most widely used anthelminthics worldwide and a mainstay of the mass drug administration programs to eliminate onchocerciasis and lymphatic filariasis in Africa. After trials in the early 1980s demonstrated the safety and efficacy of a single dose of ivermectin in reducing microfilarial levels in the skin and eye in patients with onchocerciasis, its use was investigated in patients with loiasis living in an area endemic for onchocerciasis (Richard-Lenoble et al. 1988). In two separate studies, a total of 35 patients with low-microfilarial levels (<700 microfilariae/mL of blood) were treated with ivermectin (5–200 mcg/kg in a single dose). Significant reduction of microfilarial levels was seen with doses above 100 mcg/kg and was maximal (12% of pretreatment levels at 7 days) with the highest dose studied. The medication was generally well-tolerated, although mild to moderate posttreatment reactions (pruritus, rash, arthralgia) were observed. These studies paved the way for the use of ivermectin in mass drug elimination programs for onchocerciasis in areas coendemic for loiasis.

On the 2021 WHO Model List of Essential Medicines as an intestinal anthelminthic, an antifilarial, and a medicine for ectoparasitic infections, ivermectin (Stromectol®; Merck & Co., Inc) is approved by the United States Food and Drug Administration (US FDA) for use in adults and children >15 kg for the treatment of intestinal helminths, onchocerciasis, and scabies infestation. Ivermectin is considered Pregnancy Category C because of teratogenic effects in rodents. Although a recent systematic review and meta-analysis found no clear evidence of adverse outcomes related to inadvertent ivermectin exposure in pregnancy in humans, the authors concluded that there is insufficient evidence to support its use in pregnant women at this time (Nicolas et al. 2020).

2.2.1 Mechanism of Action

Ivermectin binds to glutamate-gated chloride ion channels in invertebrate nerve and muscle cells. It has microfilaricidal activity against filarial parasites *in vivo*, including *Loa loa*, but little to no effect on adult worms. Like DEC, ivermectin does not appear to be directly toxic to microfilariae, including *Loa loa* microfilariae, *in vitro* (Njouendou et al. 2018). It has, however, been shown to affect the excretory-secretory apparatus in *B. malayi* leading to reduced secretion of proteins and extracellular vesicles (Moreno et al. 2010; Loghry et al. 2020). These findings have led to the hypothesis that ivermectin acts by impairing release of immunomodulatory proteins by the parasite exposing it to host immune attack. A recent study in healthy volunteers failed to demonstrate altered serum cytokine levels or immune cell activation following a single dose of ivermectin (Wilson et al. 2021), consistent with this hypothesis.

2.2.2 Efficacy in Loiasis

Following the initial trials of ivermectin in Gabonese patients with low levels of *L. loa* microfilariae (Richard-Lenoble et al. 1988), several small clinical trials explored the use of single-dose ivermectin (200 mcg/kg) in patients with moderate to high levels of *L. loa* microfilaremia (Carme et al. 1991b; Chippaux et al. 1992) (Table 3). These studies confirmed the incomplete clearance of *Loa loa* microfilariae with ivermectin and noted an increase in the frequency (but not severity) of adverse effects and eosinophilia in patients with higher microfilarial loads. Repeated dosing (Chippaux et al. 1992) or higher dose ivermectin (300–400 mcg/kg) (Martin-Prevel et al. 1993) improved microfilarial suppression without a concomitant increase in severe adverse reactions. Subsequent clinical trials have generally supported these findings (Table 3), as illustrated by a systematic review and meta-analysis that demonstrated an overall reduction of microfilarial levels by 90% at 3 weeks after single-dose ivermectin therapy (150–200 mcg/kg) with continued suppression for up to 1 year (Pion et al. 2019). Of note, most study participants (75.4%) continued to have detectable microfilaremia 1 year after treatment, although the number of participants with >20,000 microfilariae/mL decreased from 16.5% to 0.3%. These results are similar to those from a follow-up study in the context of Test and not Treat in which only two of 6983 individuals with microfilarial levels <25,000/mL (<0.1%) treated with ivermectin 2 years earlier were found to have microfilarial levels above 20,000/mL (Pion et al. 2020). Whereas these data do not circumvent the safety issues associated with first-time ivermectin administration to patients with very high levels of *L. loa* microfilaremia, they do suggest that repeat administration of ivermectin within 1–2 years of a prior dose is likely safe.

Although reports of fatal encephalopathy in patients with loiasis who received ivermectin as part of mass drug administration programs led to interruption of ivermectin treatment in some regions, community-based yearly ivermectin administration has continued in most areas hyper- and meso-endemic for onchocerciasis irrespective of *L. loa* endemicity. This has permitted analysis of the effects of repeated ivermectin dosing on *L. loa* prevalence and microfilaremia in relatively large populations (Ranque et al. 1996; Wanji et al. 2018). As might be expected given the lack of macrofilaricidal activity, community *L. loa* microfilarial levels tend to decrease with repeated ivermectin treatment, although persistence of *L. loa* transmission has consistently been observed despite annual distribution of ivermectin for many years (Kouam et al. 2013; Wanji et al. 2018; Nana-Djeunga et al. 2020). In one large longitudinal cross-sectional study in Cameroon where baseline data prior to ivermectin distribution were available, prevalence of microfilaremia decreased after 9–14 years of annual ivermectin distribution in only one of two sites with small numbers of individuals identified at both sites with >30,000 microfilariae/mL of blood (Wanji et al. 2018). Noncompliance with annual treatment was a major risk factor for persistence of high-microfilarial load.

Table 3 Efficacy of ivermectin monotherapy in reducing blood microfilarial levels in loiasis

Reference	No. participants Study location Age range	Microfilarial levels per mL blood (range)	Study design	Dose (mcg/kg)	Dosing frequency	Reduction in microfilaremia	Adverse events
Community-based studies in endemic areas							
Pion et al. (2020)	$n = 6983$ Cameroon ≥5 years	Geometric mean 2550 (0–25,000)	Longitudinal	150	Single dose	Reduction in overall prevalence from 17.3% to 13% with >50% reduction in microfilarial intensity after 18 months	Mild to moderate; no serious adverse events (Kamgno et al. 2017)
Nana-Djeunga et al. (2020)	$n = 376$ Cameroon ≥5 years	Mean 839	Repeated cross-sectional	150	Yearly for 18 years	Reduction in overall prevalence from 12.4% to 3.7% with 99.8% reduction in overall microfilarial intensity	NA
Wanji et al. (2018)	$n = 2261$ (excludes sites without baseline data) Cameroon ≥10 years	NA	Repeated cross-sectional	150	Yearly for 9–14 years	Prevalence of microfilaremia decreased from 30.5% to 17.9% in one site and remained stable in the other (8.1% to 7.8%); intensity decreased in both sites	NA
Gardon et al. (1997b)	$n = 533$ Cameroon >15 years	Geometric mean 503.8 (0–198,660)	Longitudinal	150	Yearly for 2 years	73% at 1 year and >74% at 2 years	NA
Ranque et al. (1996)	$n = 448$ Cameroon ≥2 years	Geometric mean 8813 (67–101,690)	Repeated cross-sectional	200	Every 3 months for 2 years	Geometric mean reduction of 92% at 2 years with 80% clearance rate in microfilaria-positive patients	Severe reactions in patients with high-microfilarial levels

Prospective clinical trials in endemic areas							
Kamgno et al. (2007)	$n = 95$ Cameroon ≥15 years	(100–25,000)	Randomized double-blind	25 vs. 150	Single dose	51.7% in low dose at day 3 compared to 64.1% in standard dose ($p < 0.01$)	Mild to moderate in <50%
Kombila et al. (1998)	$n = 109$ Gabon ≥5 years	NA	Open label	200	Monthly for 6 months	>99% average reduction in microfilarial load and prevalence at 12 months	Severe in only one patient with microfilarial load >10,000/mL
Duong et al. (1997)	$n = 71$ Gabon >15 years	Geometric mean 38.8 (0.25–35,700)	Open label	200	Single dose	88.6% at 10–12 months with 63% clearance rate	Mild to moderate in 50%
Ducorps et al. (1995)	$n = 112$ Cameroon >18 years	Geometric mean 15,725 (633–164,550)	Open label	200	Single dose	NA	Mild to moderate reactions in 71%; one nonfatal neurologic complication
Martin-Prevel et al. (1993)	$n = 31$ Gabon ≥5 years	(7–7700)	Open label	300–400	Single dose	91% for up to 3 months	Mild to moderate in 64% (significantly increased incidence in higher dose group)
Chippaux et al. (1992)	$n = 110$ Cameroon >5 years	NA	Open label	200	Every 3 months for 2–3 doses	70% for up to 3 months	Mild to moderate in 45% (more severe with high microfilaremia)
Carme et al. (1991b)	$n = 28$ Republic of Congo NA	2500–46,000	Open label	200	Single dose	90% for up to 14 days	Mild to moderate in 30%
Richard-Lenoble et al. (1988)	$n = 35$ Gabon ≥13 years	NA	Open label	5–200	Single dose	88% for up to 28 days (in 200 mcg/kg group)	Mild to moderate in 59% receiving highest dose

(continued)

Table 3 (continued)

Reference	No. participants Study location Age range	Microfilarial levels per mL blood (range)	Study design	Dose (mcg/kg)	Dosing frequency	Reduction in microfilaremia	Adverse events
Retrospective analyses in immigrants and travelers							
Bouchaud et al. (2020)	$n = 69$ varied NA	NA	Multicenter	200	1–6 doses	Resolution of symptoms and clearance of microfilaremia in 52%	NA
Gobbi et al. (2018)	$n = 39$ varied NA	Median 4820	Multicenter	150–200	Single dose	Resolution of symptoms in 52% and microfilarial clearance and resolution of eosinophilia in 17.9%	NA

Note: Only studies with >20 participants are included

Ivermectin has been used with some success in combination with macrofilaricidal therapy with DEC (Table 2) or albendazole (Arrey-Agbor et al. 2018; Gobbi et al. 2018, 2019). In the case of combination therapy with DEC, data are conflicting but seem to suggest an improvement in cure rate (resolution of clinical symptoms and microfilaremia) in patients who also received a dose of ivermectin (Table 2). Since a reliable microfilarial threshold for serious adverse events has been determined for ivermectin (but not for DEC), pretreatment with ivermectin provides a theoretical safety advantage in patients with <20,000 microfilariae/mL of blood compared to DEC treatment alone. In a retrospective case series of 16 patients with loiasis treated with albendazole (400 mg twice daily) for 28 days followed by single-dose ivermectin (200 mcg/kg), 15 of 16 patients had resolution of symptoms and eosinophilia at 6 months posttreatment (Gobbi et al. 2019). One patient required a second dose of ivermectin at 4 months for persistent low-level microfilaremia and a second patient had recurrent Calabar swellings 1 month after treatment and received a course of DEC with resolution. Patients were followed for a median of 3.5 years (range 14 months to 12 years) without recurrence. Other studies have confirmed these results and suggest that ivermectin alone may be sufficient to induce remission of clinical symptoms (Bouchaud et al. 2020).

2.2.3 Safety Concerns

As with DEC, mild to moderate side effects and peripheral eosinophilia related to microfilarial killing are common following ivermectin therapy, and there is a risk of severe neurologic events in patients with high-microfilarial levels. Although decreasing the dose of ivermectin does reduce the average microfilarial clearance at 3 and 7 days in patients with loiasis (Richard-Lenoble et al. 1988; Kamgno et al. 2000, 2007), the magnitude of the reduction is highly variable between individuals and insufficient to justify the use of low-dose therapy to prevent serious posttreatment complications (Kamgno et al. 2007).

Excluding effects related to its microfilaricidal activity (see "Complications of treatment" above), ivermectin is extremely well-tolerated at the doses used for anthelmintic therapy. There is one published case of probable ivermectin-induced hepatitis in a patient with loiasis (Veit et al. 2006) and possible or probable ivermectin-associated hepatotoxicity has been reported in patients receiving ivermectin for scabies or Sars-CoV-2 infection (Boudou et al. 2020; Oscanoa et al. 2022). In addition, neurologic, gastrointestinal, and musculoskeletal toxicity, including at least one fatality, have been reported in patients taking high doses for the prevention or treatment of COVID-19 (Hoang et al. 2022). Unlike diethylcarbamazine, ivermectin does not provoke severe Mazzotti reactions and is safe for use in patients coinfected with onchocerciasis.

2.3 Albendazole

The benzimidazole derivative, albendazole, is a broad-spectrum anthelminthic with activity against a wide variety of nematodes and cestodes. Developed in 1975 by

SmithKline Corporation as a veterinary product, early studies demonstrated efficacy against adult *Brugia* parasites with little or no direct effect on microfilariae (Denham et al. 1980) prompting clinical studies in onchocerciasis (Awadzi et al. 1991; Cline et al. 1992). Although the benzimidazole, mebendazole, had been used with some success in the treatment of loiasis, poor oral availability and side effects of the high doses required for antifilarial activity limited its utility (Richard-Lenoble et al. 1985; Van Hoegaerden and Flocard 1985, 1986; Hoegaerden et al. 1987). Albendazole, which is better absorbed and extremely well-tolerated in the prior clinical trials in onchocerciasis, provided a logical alternative.

Albendazole is on the 2021 WHO List of Essential Medicines and Teratogenic in rats and rabbits, and has been given safely to pregnant women, mostly as a single dose in early second trimester in the context of soil-transmitted helminth deworming programs (Bradley and Horton 2001). Several small studies have examined the consequences of albendazole administration in the first trimester of pregnancy (Bradley and Horton 2001), including one study in Ghana, which found no evidence of increased risk of congenital malformation or spontaneous abortion in 50 women who inadvertently received single-dose albendazole during mass drug administration for filariasis control compared to 293 women who were excluded from treatment (Gyapong et al. 2003). Nevertheless, data are insufficient to support the routine use of albendazole in pregnant women with loiasis. Albendazole is also not recommended for children less than 2 years of age.

2.3.1 Mechanism of Action

As is true for all benzimidazole, albendazole (methyl 5-(propylthio)-2-benzimidazolecarbamate), selectively binds parasite tubulin leading to impaired embryogenesis and death of adult worms. It has little to no direct effect on microfilariae but leads to a reduction in microfilarial counts over time because of its macrofilaricidal and embryotoxic effects. Albendazole is variably absorbed from the gastrointestinal tract due to low aqueous solubility. However, levels of its active metabolite, albendazole sulfoxide, are increased up to fivefold when albendazole is administered with a fatty meal (Whittaker et al. 2022).

2.3.2 Efficacy in Loiasis

The first clinical trial of albendazole in patients with loiasis was a double-blind, placebo-controlled trial designed to assess drug safety and the effect on blood microfilarial levels. Twenty-three men with microfilarial levels ranging from 100 to 30,000/mL were randomly assigned to albendazole (200 mg twice daily for 21 days) or placebo (Klion et al. 1993). The drug was extremely well-tolerated and led to a slow reduction in blood microfilarial levels beginning at day 14, consistent with a primary effect on adult worms, with 80% reduction in geometric mean microfilarial levels achieved at 6 months. These results led to multiple trials of albendazole in loiasis with the primary goal of finding a more practical regimen that could be used to safely lower microfilarial loads without the need for monitoring of liver function and hematologic parameters. Unfortunately, shorter courses of albendazole have been less effective in lowering microfilaremia even when given at higher dose or repeated monthly for up to 6 months (Kamgno and Boussinesq 2002; Tsague-Dongmo et al. 2002; Tabi et al. 2004; Kamgno et al. 2016) (Table 4).

Several studies suggest that albendazole has the potential to cure *L. loa* infection. In one early study, albendazole therapy (200 mg orally twice daily for 3 weeks) led to long-term resolution of symptoms in three amicrofilaremic patients with symptomatic loiasis who had failed multiple courses of DEC (Klion et al. 1999). In a second long-term follow-up study, resolution of symptoms, microfilaremia, and eosinophilia were reported for up to 12 years in 14/16 patients who received albendazole (400 mg orally twice daily for 28 days) in combination with single-dose ivermectin (Gobbi et al. 2019). Although cure rates with albendazole, alone or in combination with DEC or ivermectin, have ranged from 33% to 100% in other retrospective series of patients with loiasis (Gantois et al. 2013; Saito et al. 2015; Gobbi et al. 2018; Bottieau et al. 2022), the relatively small numbers of patients treated, variable dosing, and differences in duration of follow-up preclude definitive conclusions.

2.3.3 Safety Concerns

Known side effects of albendazole include gastrointestinal symptoms, liver function abnormalities, and hypersensitivity reactions, including Stevens-Johnson syndrome. Rare, but potentially fatal, toxicities include bone marrow suppression, aplastic anemia, and agranulocytosis, all of which are more common in people with underlying liver disease. Two cases of encephalopathy temporally related to albendazole treatment of loiasis have been reported (Blum et al. 2001; Volpicelli et al. 2020; Metais et al. 2021). Pretreatment microfilarial levels were 152/mL in one patient and 15,000/mL in the second. Extensive evaluations revealed no alternative etiology for the events. Although these cases suggest that albendazole can be associated with neurologic events in patients with loiasis, the widespread use of this drug for soil-transmitted helminth infection in areas endemic for loiasis without reports of encephalopathy is evidence for the rarity of this complication. Albendazole can provoke an inflammatory response around lesions of cysticercosis with exacerbation of neurologic symptoms and should be used in caution in patients with a history of unexplained seizures or potential exposure to the pork tapeworm, *Taenia solium* (Ramos-Zúñiga et al. 2013).

2.4 Doxycycline

Doxycycline is a semisynthetic tetracycline antibiotic with activity against a wide variety of microorganisms, including bacteria, rickettsia, and protozoa. Multiple studies have demonstrated that *Loa loa* parasites do not harbor the intracellular endosymbiont *Wolbachia*, the target of doxycycline therapy in other filarial species (McGarry et al. 2003; Büttner et al. 2003; Grobusch et al. 2003; Desjardins et al. 2013), and the expected lack of efficacy of doxycycline treatment was confirmed in two patients with loiasis treated for 6 weeks without appreciable effect on blood microfilarial levels (Brouqui et al. 2001).

Table 4 Reduction in microfilaremia with albendazole

Reference	Study population	Study design	Dose	Duration of treatment	Endpoint	Outcome
Kamgno et al. (2016)	Endemic ($n = 60$)	Randomized placebo-controlled[a]	800 mg	Single dose every 2 months for 2 or 6 doses	Mf count up to 24 months	Reduction in mf count by ≥50% in 50% of subjects in the 6 dose group but 0/9 participants with >30,000 mf/mL pretreatment had reduction below 8000 mf/mL
Tsague-Dongmo et al. (2002)	Endemic ($n = 95$)	Randomized comparison to multivitamin[a]	400 mg twice daily	3 days	Mf count up to 9 months	No significant reduction at any time point
Kamgno and Boussinesq (2002)	Endemic ($n = 45$)	Randomized comparison to mebendazole[a]	600 mg	Single dose	Mf count up to 10 months	Variable reduction at 300 days (0 to tenfold)
Tabi et al. (2004)	Endemic ($n = 99$)	Randomized placebo-controlled crossover[a]	400 mg daily	3 days	Mf count at 6 months	Transient decrease at 3 months
Klion et al. (1993)	Endemic ($n = 23$)	Randomized, double-blind, placebo-controlled	200 mg twice daily	21 days	Mf count up to 6 months	Reduction in microfilaremia count by 80% in albendazole group at 6 months

[a]Groups were stratified by microfilarial count for randomization

2.5 Other Agents in Clinical Development

With the expansion of mass drug treatment programs for onchocerciasis in Africa and the risk of ivermectin-related serious adverse events, there has been renewed interest in the identification of alternative agents for the safe reduction of *L. loa* microfilaremia (Table 5). To this end, *in vitro* culture methods for maintenance of *L. loa* microfilariae (Zofou et al. 2018) and immunodeficient mouse models of loiasis that mimic the subcutaneous migration of adult worms and microfilaremia seen in human infections and respond to microfilaricidal therapy with upregulation of type 2 inflammatory responses and eosinophilia have been developed (Pionnier et al. 2019). Whereas many alternative agents, including plant extracts (Mengome et al. 2010; Ngwewondo et al. 2018), have demonstrated activity against *L. loa* parasites *in vitro*, only a few candidate drugs for loiasis have been tested *in vivo*. Most are commercially available drugs used to treat other disorders or veterinary anthelmintics repurposed for use in humans.

1. Alternative Benzimidazoles

Flubendazole and oxfendazole are benzimidazole drugs related to albendazole but with better bioavailability, activity against adult filarial worms in a variety of animal models, and no direct effect on *L. loa* microfilariae *in vitro* (O'Neill et al. 2018) or in animal models (Pionnier et al. 2019). Although clinical development of flubendazole was halted due to teratogenicity and aneugenicity in preclinical trials (Lachau-Durand et al. 2019), phase I trials of oxfendazole (up to 15 mg/kg for 5 days) have been completed (An et al. 2019; Bach et al. 2020), and a phase 2 trial in onchocerciasis is planned (Pfarr et al. 2023). As with albendazole, consumption of food increased plasma levels of oxfendazole. Approximately 25% of participants in the phase 1 trials experienced a mild to moderate adverse event, of which gastrointestinal symptoms and transient laboratory abnormalities were most common. Based on the available data, oxfendazole is likely to be at least as effective as albendazole for the treatment of loiasis with a comparable safety profile due to the lack of effect on microfilariae.

2. Emodepside

Emodepside is a semisynthetic bismorpholino-cyclooctadepsipeptide derived from the fungus *Rossellinia*. Studies in *C. elegans* and soil-transmitted helminths suggest that it causes paralysis and death through dual binding of the nematode receptor, latrophilin, and the SLO-1 ion channel (Krücken et al. 2012). Although emodepside has activity against all stages of filarial parasites, including microfilariae, stage- and species-specific differences in activity have been described *in vitro* and *in vivo* (Kashyap et al. 2019; Bah et al. 2021; Hübner et al. 2021). In one study comparing the effects of emodepside on different filarial parasites *in vitro*, adult worms of all filarial species tested (except for *Brugia pahangi*) were more susceptible to emodepside killing than other life cycle stages, suggesting that lower

Table 5 Agents in development for the treatment of filariasis

	Oxfendazole	Emodepside	Levamisole	Imatinib
Manufacturer	Drugs for Neglected Diseases Initiative	Bayer Health Care Group	Multiple	Multiple
Class	Benzimidazole	Depsipeptide	Nicotinic acetylcholine receptor agonist	Tyrosine kinase inhibitor
Formula	$C_{15}H_{13}N_3 0_3 S$	$C_{60}H_{90}N_6 0_{14}$	$C_{11}H_{12}N_2S$	$C_{29}H_{31}N_7 0$
Structure	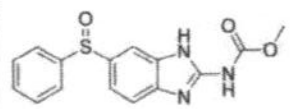	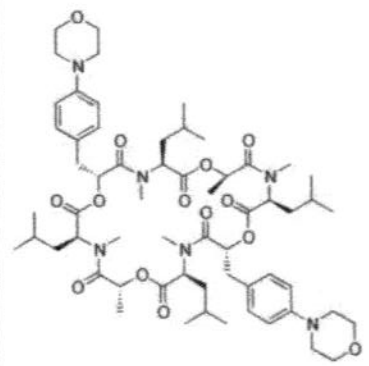	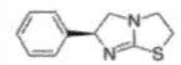	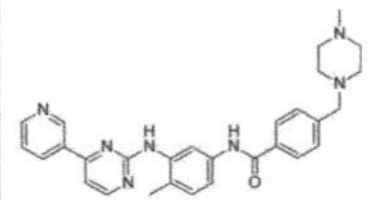
Mechanism of action	Binds parasite tubulin leading to impaired embryogenesis and death of adult worms	Binds to G-protein coupled receptors called latrophilins and the Ca2+-gated K+ channel SLO-1 inducing neuromuscular paralysis	Binds to nicotinic acetylcholine receptor inducing muscular dysfunction	Irreversibly binds to enzyme active site preventing protein phosphorylation and inducing cellular apoptosis
Side effect profile	Generally well-tolerated; side effect profile likely similar to albendazole	Generally well-tolerated; mild to moderate, transient visual and neurologic symptoms reported	Generally well-tolerated; rare agranulocytosis and/or vasculitis reported	Generally well-tolerated; liver toxicity and dose-related cytopenias each reported in up to 5% of patients
Antifilarial activity	Adult parasites	All parasite stages; possible preferential effect on adult parasites	Microfilaricidal; variable activity on adult worms	All parasite stages
Data in loiasis	No effect on *Loa loa* microfilariae in vitro or in mouse model; no data on adult *Loa loa* worms	No data available	Transient lowering of microfilarial levels in humans after single dose	No effect in pilot placebo-controlled trial ($n = 20$)
Current indications	Veterinary: prophylaxis and treatment of nematodes	Veterinary: prophylaxis and treatment of nematodes	Human: treatment of intestinal helminths	Human: treatment of multiple myeloid and nonmyeloid malignancies

doses may preferentially target adult worms (Hübner et al. 2021). Unfortunately, this preferential effect on adult worms was not reproduced in a study of emodepside treatment of cattle infected with *Onchocerca ochengi* (Bah et al. 2021). Of note, there have been no studies published to date assessing the effect of emodepside on *L. loa* parasites.

Emodepside has been administered to humans. No major safety concerns were identified in phase 1 ascending dose trials of emodepside in 103 healthy male volunteers (maximum dose 10 mg orally twice daily for 10 days), although transient, mild, drug-related visual and neurologic disturbances were noted in approximately 45% of the study subjects (Gillon et al. 2021). More recently, emodepside was administered to a total of 328 participants in two randomized controlled trials of emodepside for the treatment of soil-transmitted helminth infection (Mrimi et al. 2023). Emodepside was effective against both *Trichuris* and hookworm infection and was generally well-tolerated, although transient mild to moderate headache, blurred vision, and/or dizziness were reported in a substantial number of participants. A phase 2 trial of emodepside for the treatment of onchocerciasis is ongoing with plans to assess the effects of treatment on adult worm viability and microfilarial clearance compared to ivermectin (NCT05180461). The results of this trial should provide additional information to help determine whether emodepside might be of use for the treatment of loiasis.

3. Levamisole

Levamisole is a nicotinic acetylcholine receptor antagonist with activity against a broad range of helminths, including filariae. In animal models of filariasis, levamisole is a potent microfilaricide with variable efficacy against adult worms that appears to be dose- and filarial species-related (Zahner and Schares 1993). Although levamisole is on the World Health Organization's list of Essential Medicines and has been used for decades for the treatment of intestinal helminths, it causes agranulocytosis in up to 5% of people and is associated with an unusual form of vasculitis seen primarily in users of levamisole-adulterated cocaine (Lee et al. 2012). Fatalities due to these complications led to its removal from the market in the US and Canada in 2000 and 2003, respectively. Prior clinical trials of levamisole (30–300 mg for up to 21 days) in patients with onchocerciasis or lymphatic filariasis showed a moderate short-lasting effect on microfilarial levels in most, but not all, trials (reviewed in Campillo et al. 2022), and, when reported, adverse events were generally mild and temporally related to the decrease in microfilarial counts. A randomized, placebo-controlled, double-blind ascending single-dose trial of levamisole for the treatment of loiasis was conducted in 2022 (Campillo et al. 2022). A total of 224 individuals with microfilaremic loiasis (5–69,085 mf/mL blood) received levamisole at doses ranging from 1.0 to 2.5 mg/kg. Although median microfilarial count was significantly reduced in the highest dose levamisole group compared to placebo at days 2,7, and 30, there was considerable interindividual variation and the proportion of patients with at least a 40% reduction in microfilarial counts was only different between the two groups at day 2. Mild adverse events, including vomiting and dizziness, were significantly more frequent in the group that received levamisole and in patients with higher microfilarial levels. No serious adverse events were reported. Given the rapid effect on *L. loa* microfilarial levels and the association of this effect with adverse events, it is unclear whether levamisole provides any clear advantage over DEC for the treatment of loiasis.

4. Imatinib

Imatinib is a selective tyrosine kinase inhibitor with activity against c-Abl and platelet-derived growth factors alpha and beta. Originally developed for the treatment of Philadelphia chromosome-positive chronic myelogenous leukemia, imatinib has been given to millions of people worldwide. Although imatinib is generally well-tolerated, mild to moderate edema and gastrointestinal complaints are common and more severe side effects, including hepatotoxicity and dose-related severe cytopenia, occur in up to 5% of patients with short-term use. Rare cases of cardiac necrosis have also been reported in patients with preexisting eosinophilic myocarditis. Imatinib is approved for use in children older than 2 years but is teratogenic and should not be administered to pregnant women. Following the identification of an Abl homolog during the sequencing of the *L. loa* genome, *in vitro* studies confirmed the activity of imatinib against all life cycle stages of *Brugia malayi* (O'Connell et al. 2015). Although administration of a single dose of imatinib (600 mg orally) to a patient with loiasis resulted in a slow decrease in the blood microfilarial count over 6 days from 2250 to 150 mf/mL of blood with minimal side effects and improvement in symptoms (O'Connell and Nutman 2017), these findings were not confirmed in a subsequent placebo-controlled clinical trial and two of the 15 patients who received imatinib developed grade 3 neutropenia (absolute neutrophil count 0.5–1.0 × 10^9/L; results available at clinicaltrials.gov; NCT02644525).

3 Adjuncts to Chemotherapy

3.1 Glucocorticoids

Pretreatment with prednisone has been shown to increase the number of residual microfilariae following DEC treatment in patients with lymphatic filariasis (Schofield and Rowley 1961) and onchocerciasis (Stingl et al. 1988). Although pretreatment with glucocorticoids has been used successfully to reduce allergic symptoms (i.e., angioedema, urticaria) related to microfilarial killing in loiasis (Thompson 1956), there is substantial evidence from human cases reports to suggest that this approach is ineffective for the prevention of serious complications of therapy, including encephalopathy (Carme et al. 1991a) and renal failure. Moreover, in a study in baboons experimentally infected with *L. loa* and treated with ivermectin, administration of prednisone was associated with increased numbers of microfilariae in the brain and kidneys (Wanji et al. 2017). That said, short courses of glucocorticoids do not appear to reduce the likelihood of cure following a single course of DEC treatment and have been used historically to reduce inflammatory complications (Herrick et al. 2015).

3.2 Antihistamines

Antihistamine therapy is often administered for prevention or treatment of allergic symptoms (e.g., urticaria, angioedema, pruritus) following microfilaricidal therapy in patients with loiasis. Although antihistamines appear to be moderately effective in reducing the intensity of these posttreatment symptoms(Carme et al. 1982; Bouchaud et al. 2020), they do not prevent severe posttreatment sequelae.

3.3 Apheresis

The use of cytapheresis to lower *L. loa* microfilarial levels prior to administration of DEC was first described in 1983 by Muylle et al. (1983). The authors described two patients with loiasis and moderate levels of microfilaremia (9000/mL and 3750/mL) who were intolerant of DEC due to severe side effects. Three apheresis sessions were conducted per patient with variable reduction in microfilarial counts but improved tolerance of DEC therapy. Since that time, there have been numerous reports of the successful use of apheresis to reduce DEC-related adverse events in patients with loiasis (Abel et al. 1986; Chandenier et al. 1987; Norgan et al. 2018). A recent systematic review of the use of apheresis in loiasis identified 14 publications describing 34 patients (Odedra et al. 2019). The median % reduction of microfilariae was 51.7%, and nine of the 12 patients with a known baseline microfilarial load of >8000/mL had reduction below this level prior to DEC treatment. The procedure was well-tolerated, and DEC-related adverse events' postapheresis was seen in only three patients (8.8%). Despite modifications to simply the apheresis procedure and reduce concomitant depletion of platelets (the most common adverse event in early studies) (Norgan et al. 2018; Odedra et al. 2019), cost and access to a specialized apheresis unit remain major limitations to the widespread use of this technique to lower microfilarial counts prior to definitive treatment.

4 Treatment Approach

Given recent data demonstrating an increased mortality risk in patients with microfilaremic loiasis and the potential for rare but clinically significant spontaneous complications of *L. loa* infection, including fatal encephalopathy, renal failure, and endomyocardial fibrosis, benefits of treatment are clear. However, in the absence of a safe and effective curative regimen, individualization of therapy is necessary. Moreover, the decision to treat and choice of therapeutic agent(s) depend on many different factors, including the severity of clinical manifestations, degree of blood eosinophilia, presence and intensity of microfilaremia, concomitant infections, reinfection risk and access to drugs, and/or posttreatment monitoring.

The literature provides little guidance on treatment of loiasis. The current Center for Disease Control guidelines recommend DEC treatment (8–10 mg/kg daily for 21 days) for patients with microfilarial levels less than 8000/mL and provide no options for infected patients with microfilarial levels above this threshold (https://www.cdc.gov/parasites/loiasis/health_professionals/index.html). The World Health Organization provides no recommendations for the treatment of loiasis on its website but comments that patients with high levels of microfilaremia can develop severe adverse events if treated with ivermectin (https://espen.afro.who.int/diseases/loiasis). Finally, a PubMed search identified a single publication from 2012 outlining an approach to treatment based on microfilarial levels (Boussinesq 2012). Recognizing the lack of data supporting the safety cutoffs for DEC treatment and the paucity of well-controlled prospective treatment studies in patients with loiasis, a potential treatment algorithm based in part on the abovementioned treatment approach is provided in Fig. 1.

For patients with <2000 microfilariae/mL, DEC is the treatment of choice for patients residing outside of an endemic area because of the high likelihood of definitive cure. If there is a risk of reinfection, the relative benefit of curative therapy should be weighed against the risk of treatment side effects, such as pruritus, angioedema, musculoskeletal complaints, and fever, which occur in approximately 50% of patients. Although most patients with low or undetectable microfilarial levels can be treated as outpatients, rare cases of significant complications have been reported so access to medical care should also factor into the decision to treat with DEC. Short courses (2–3 days) of glucocorticoids or oral antihistamines can be used to reduce allergic symptoms, such as angioedema and urticaria, and do not appear to affect treatment efficacy. If DEC is unavailable or contraindicated, albendazole with or without concomitant ivermectin may be effective. Finally, albendazole (200 mg twice daily for 21 days) should be considered in patients who have persistent symptoms despite multiple courses of DEC.

For patients with 2000–20,000 microfilariae/mL, pretreatment with ivermectin is advised to lower the microfilarial load before definitive treatment with DEC as above. Substantial data support the safety of ivermectin treatment for patients with microfilarial loads in this range, although moderately severe adverse events do occur and may require referral to a health care provider. Repeated doses of ivermectin may be necessary at 1–3 month intervals to achieve levels <2000 microfilariae/mL. Albendazole (200 mg twice daily for 21 days) can also be used to lower microfilarial levels, although peak effect requires 6 months.

In cases of microfilaremia >20,000/mL, neither DEC nor ivermectin is without risk of severe complications, including potentially fatal encephalopathy. Consequently, reduction of microfilarial levels is recommended. Whereas apheresis has been used successfully for this purpose, this is not practical in most situations. Thus, until better options become available, albendazole (200 mg twice daily for 21 days) is the treatment of choice for patients with extremely high-microfilarial loads.

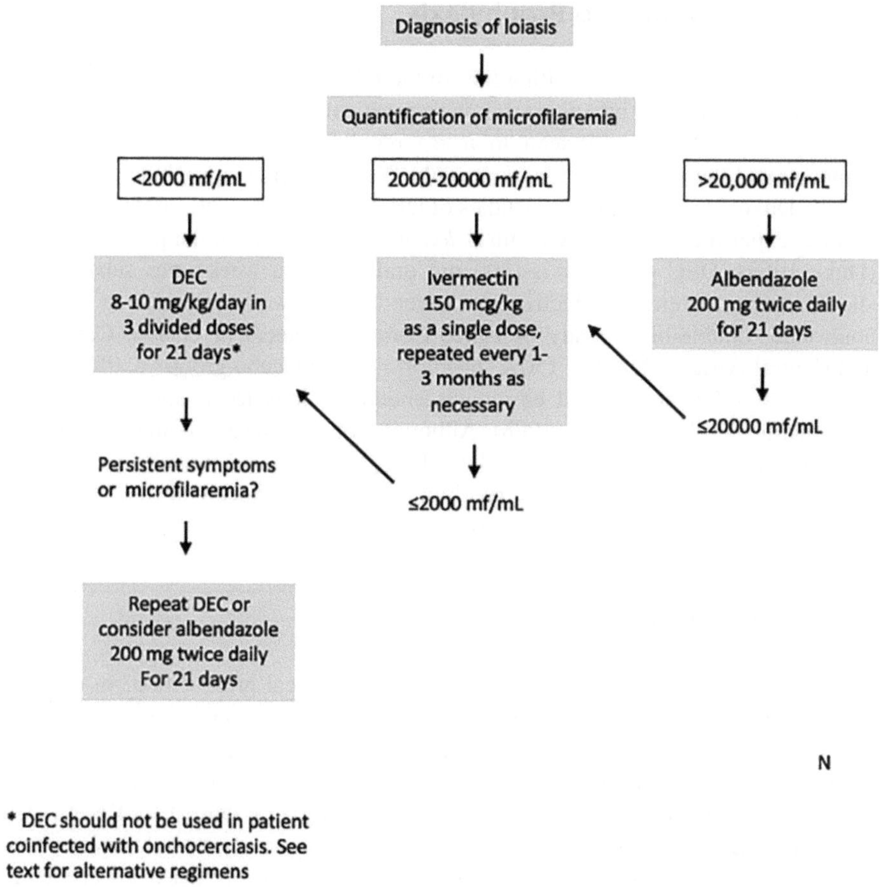

Fig. 1 A stepwise approach to the treatment of loiasis. A simple approach to the treatment of loiasis based on microfilarial density is illustrated. For alternatives and additional considerations, please see main text

4.1 Coinfection with Onchocerciasis

DEC should not be used in patients with known or suspected onchocerciasis. Pretreatment with ivermectin (150 mcg/kg as a single dose) may be useful in this context, although the optimal time to wait before DEC can be safely administered has not been established. Alternatively, albendazole therapy (200 mg twice daily for 3 weeks) can be used with ivermectin to treat coinfected patients.

5 Prevention and Prophylaxis

Loiasis is transmitted by day-biting flies of the *Chrysops* genus. Whereas avoidance of the vector prevents infection, this is impractical in most situations. Moreover, *Chrysops* are relatively resistant to insect repellents. After studies in mandrills demonstrated the efficacy of low-dose daily DEC as a chemoprophylactic agent in loiasis (Duke 1961), preliminary studies of DEC prophylaxis were conducted in four humans experimentally infected with *L. loa* and suggested a similar protective effect (Duke 1963). DEC prophylaxis (300 mg orally once a week) was subsequently shown to be extremely effective in preventing *L. loa* infection in a placebo-controlled, double-blind study of Peace Corps volunteers in Gabon, Cameroon, and Central African Republic (30% infection rate in placebo group vs. 0% in DEC group, $p < 0.02$) and should be recommended for long-term travelers to highly endemic areas (Nutman et al. 1988). Although noncompliance, defined as missing more than one dose per month was relatively high at 47%, this was attributed to side effects in only three subjects (one receiving placebo two receiving DEC) and was similar between the two study groups.

6 Conclusions

Loiasis is a complex infection with variability in clinical presentation. In the past, treatment was reserved for individuals with significant symptoms or extremely high-eosinophil counts due to the potential risk of neurologic and other severe complications of treatment in patients with high levels of microfilaremia. As a result, infected residents of endemic areas were frequently left untreated. Recent data demonstrating excess mortality related to the presence of *Loa loa* microfilaremia irrespective of symptoms or eosinophilia have made this approach unacceptable. Although current treatment options using DEC, ivermectin, and albendazole, are imperfect, treatment of loiasis is possible even in patients with high levels of circulating microfilariae. A better understanding of the mechanisms underlying severe posttreatment reactions in loiasis and the development of new drugs that can be used to safely treat all infected patients remain unmet research needs in this regard.

Acknowledgments I would like to thank Michel Boussinesq for his helpful comments. This work was funded by the Intramural Research Program of NIAID, NIH.

References

Abel L, Ioly V, Jeni P et al (1986) Apheresis in the management of loiasis with high microfilariaemia and renal disease. Br Med J (Clin Res Ed) 292:24

Ackerman SJ, Kephart GM, Francis H et al (1990) Eosinophil degranulation. An immunologic determinant in the pathogenesis of the Mazzotti reaction in human onchocerciasis. J Immunol 144:3961–3969

Andy JJ, Ogunowo PO, Akpan NA et al (1998) Helminth associated hypereosinophilia and tropical endomyocardial fibrosis (EMF) in Nigeria. Acta Trop 69:127–140. https://doi.org/10.1016/s0001-706x(97)00125-3

An G, Murry DJ, Gajurel K et al (2019) Pharmacokinetics, safety, and tolerability of oxfendazole in healthy volunteers: a randomized, placebo-controlled first-in-human single-dose escalation study. Antimicrob Agents Chemother 63. https://doi.org/10.1128/AAC.02255-18

Arrey-Agbor DB, Nana-Djeunga HC, Mogoung-Wafo AE et al (2018) Case Report: Probable case of spontaneous encephalopathy due to loiasis and dramatic reduction of Loa loa microfilariaemia with prolonged repeated courses of albendazole. Am J Trop Med Hyg 99: 112–115. https://doi.org/10.4269/ajtmh.17-0664

Awadzi K, Hero M, Opoku O et al (1991) The chemotherapy of onchocerciasis. XV. Studies with albendazole. Trop Med Parasitol 42:356–360

Bach T, Galbiati S, Kennedy JK et al (2020) Pharmacokinetics, safety, and tolerability of oxfendazole in healthy adults in an open-label phase 1 multiple ascending dose and food effect study. Antimicrob Agents Chemother 64. https://doi.org/10.1128/AAC.01018-20

Bah GS, Schneckener S, Hahnel SR et al (2021) Emodepside targets SLO-1 channels of Onchocerca ochengi and induces broad anthelmintic effects in a bovine model of onchocerciasis. PLoS Pathog 17:e1009601. https://doi.org/10.1371/journal.ppat.1009601

Blum J, Wiestner A, Fuhr P, Hatz C (2001) Encephalopathy following Loa loa treatment with albendazole. Acta Trop 78:63–65. https://doi.org/10.1016/s0001-706x(00)00159-5

Bolla S, Boinpally RR, Poondru S et al (2002) Pharmacokinetics of diethylcarbamazine after single oral dose at two different times of day in human subjects. J Clin Pharmacol 42:327–331. https://doi.org/10.1177/00912700222011247

Bottieau E, Huits R, Van Den Broucke S et al (2022) Human filariasis in travelers and migrants: a retrospective 25-year analysis at the Institute of Tropical Medicine, Antwerp, Belgium. Clin Infect Dis 74:1972–1978. https://doi.org/10.1093/cid/ciab751

Bouchaud O, Matheron S, Loarec A et al (2020) Imported loiasis in France: a retrospective analysis of 167 cases with comparison between sub-Saharan and non sub-Saharan African patients. BMC Infect Dis 20:63. https://doi.org/10.1186/s12879-019-4740-6

Boudou E, Lespine A, Durrieu et al (2020) Serious ivermectin toxicity and human ABCB1 nonsense mutations. N Engl J Med 383(8):787–789. https://doi.org/10.1056/NEJMc1917344

Bourguinat C, Kamgno J, Boussinesq M et al (2010) Analysis of the mdr-1 gene in patients co-infected with Onchocerca volvulus and Loa loa who experienced a post-ivermectin serious adverse event. Am J Trop Med Hyg 83:28–32. https://doi.org/10.4269/ajtmh.2010.09-0734

Boussinesq M (2012) Loiasis: new epidemiologic insights and proposed treatment strategy. J Travel Med 19:140–143. https://doi.org/10.1111/j.1708-8305.2012.00605.x

Bradley M, Horton J (2001) Assessing the risk of benzimidazole therapy during pregnancy. Trans R Soc Trop Med Hyg 95:72–73. https://doi.org/10.1016/s0035-9203(01)90338-4

Brouqui P, Fournier PE, Raoult D (2001) Doxycycline and eradication of microfilaremia in patients with loiasis. Emerging Infect Dis 7:604–605. https://doi.org/10.3201/eid0707.010747

Buell KG, Whittaker C, Chesnais CB et al (2019) Atypical clinical manifestations of loiasis and their relevance for endemic populations. Open Forum. Infect Dis 6:ofz417. https://doi.org/10.1093/ofid/ofz417

Büttner DW, Wanji S, Bazzocchi C et al (2003) Obligatory symbiotic Wolbachia endobacteria are absent from Loa loa. Filaria J 2:10. https://doi.org/10.1186/1475-2883-2-10

Campillo JT, Bikita P, Hemilembolo M et al (2022) Safety and efficacy of levamisole in loiasis: a randomized, placebo controlled, double-blind clinical trial. Clin Infect Dis 75:19–27. https://doi.org/10.1093/cid/ciab906

Carme B, Boulesteix J, Boutes H, Puruehnce MF (1991a) Five cases of encephalitis during treatment of loiasis with diethylcarbamazine. Am J Trop Med Hyg 44:684–690. https://doi.org/10.4269/ajtmh.1991.44.684

Carme B, Danis M, Gentilini M (1982) Traitement de la filariose à Loa Loa : complications, résultats. A propos de 100 observations. Médecine et Maladies Infectieuses 13:184–188. https://doi.org/10.1016/S0399-077X(82)80063-2

Carme B, Ebikili B, Mbitsi A, Copin N (1991b) Therapeutic trial with ivermectin in loiasis with medium and high microfilaremia. Ann Soc Belg Med Trop 71:47–50

Cauchie C, Rutsaert J, Thys O et al (1965) Encephalitis due to Loa-Loa, treated with the combination of cortisone and carbamazine. Rev Belg Pathol Med Exp 31:232–244

Cesbron J-Y, Capron A, Vargaftig BB et al (1987) Platelets mediate the action of diethylcarbamazine on microfilariae. Nature 325:533–536

Chandenier J, Pillier-Loriette C, Datry A et al (1987) Value of cytapheresis in the treatment of loaiasis with high blood microfilaria levels. Results in 7 cases. Bull Soc Pathol Exot Filiales 80: 624–633

Chesnais CB, Pion SD, Boullé C et al (2020) Individual risk of post-ivermectin serious adverse events in subjects infected with Loa loa. EClinicalMedicine 28:100582. https://doi.org/10.1016/j.eclinm.2020.100582

Chesnais CB, Takougang I, Paguélé M et al (2017) Excess mortality associated with loiasis: a retrospective population-based cohort study. Lancet Infect Dis 17:108–116. https://doi.org/10.1016/S1473-3099(16)30405-4

Chippaux JP, Ernould JC, Gardon J et al (1992) Ivermectin treatment of loiasis. Trans R Soc Trop Med Hyg 86:289. https://doi.org/10.1016/0035-9203(92)90313-2

Churchill DR, Morris C, Fakoya A et al (1996) Clinical and laboratory features of patients with loiasis (Loa loa filariasis) in the U.K. J Infect 33:103–109. https://doi.org/10.1016/s0163-4453(96)93005-4

Cline BL, Hernandez JL, Mather FJ et al (1992) Albendazole in the treatment of onchocerciasis: double-blind clinical trial in Venezuela. Am J Trop Med Hyg 47:512–520. https://doi.org/10.4269/ajtmh.1992.47.512

Cruel T, Arborio M, Schill H et al (1997) Nephropathy and filariasis from Loa loa. Apropos of 1 case of adverse reaction to a dose of ivermectin. Bull Soc Pathol Exot 90:179–181

Datta R, Robertson A, Martin R, Kashyap S (2022) High concentrations of the anthelmintic diethylcarbamazine paralyze C. elegans independently of TRP-2. MicroPubl Biol. https://doi.org/10.17912/micropub.biology.000548

Denham DA, Liron DA, Brandt E (1980) The anthelmintic effects of albendazole on Brugia pahangi. J Helminthol 54:199–200. https://doi.org/10.1017/s0022149x00006581

Desjardins CA, Cerqueira GC, Goldberg JM et al (2013) Genomics of Loa loa, a Wolbachia-free filarial parasite of humans. Nat Genet 45:495–500. https://doi.org/10.1038/ng.2585

Ducorps M, Gardon-Wendel N, Ranque S et al (1995) Secondary effects of the treatment of hypermicrofilaremic loiasis using ivermectin. Bull Soc Pathol Exot 88:105–112

Duke BOL (1963) Studies on the chemoprophylaxis of loiasis. II. Observations on diethylcarbamazine citrate (Banocide) as a prophylactic in man. Ann Trop Med Parasitol 57:82–96

Duke BO (1990) Effect of diethylcarbamazine on adult Loa in monkeys. Ann Trop Med Parasitol 84:387–392. https://doi.org/10.1080/00034983.1990.11812484

Duke BO (1961) Studies on the chemoprophylaxis of loiasis. I. Experiments on monkeys, with special reference to diethylcarbamazine (Banocide). Ann Trop Med Parasitol 55:447–451. https://doi.org/10.1080/00034983.1961.11686073

Duong TH, Kombila M, Ferrer A et al (1997) Reduced Loa loa microfilaria count ten to twelve months after a single dose of ivermectin. Trans R Soc Trop Med Hyg 91:592–593. https://doi.org/10.1016/s0035-9203(97)90040-7

Emukah E, Rakers LJ, Kahansim B et al (2018) In Southern Nigeria Loa loa blood microfilaria density is very low even in areas with high prevalence of loiasis: results of a survey using the new LoaScope technology. Am J Trop Med Hyg 99:116–123. https://doi.org/10.4269/ajtmh.18-0163

Fobi G, Gardon J, Santiago M et al (2000) Ocular findings after ivermectin treatment of patients with high Loa loa microfilaremia. Ophthalmic Epidemiol 7:27–39

Fokom-Domgue J, Pion SD, Gounoue R et al (2014) Absence of an association between Plasmodium falciparum infection and post-ivermectin Loa-related non-neurologic serious adverse events. Am J Trop Med Hyg 90:335–338. https://doi.org/10.4269/ajtmh.13-0189

Gantois N, Rapp C, Gautret P et al (2013) Imported loiasis in France: a retrospective analysis of 47 cases. Travel Med Infect Dis 11:366–373. https://doi.org/10.1016/j.tmaid.2013.08.005

Gardon J, Gardon-Wendel N, Demanga-Ngangue et al (1997a) Serious reactions after mass treatment of onchocerciasis with ivermectin in an area endemic for Loa loa infection. Lancet 350:18–22. https://doi.org/10.1016/S0140-6736(96)11094-1

Gardon J, Kamgno J, Folefack G et al (1997b) Marked decrease in Loa loa microfilaraemia six and twelve months after a single dose of ivermectin. Trans R Soc Trop Med Hyg 91:593–594. https://doi.org/10.1016/s0035-9203(97)90041-9

Gillon J-Y, Dennison J, van den Berg F et al (2021) Safety, tolerability and pharmacokinetics of emodepside, a potential novel treatment for onchocerciasis (river blindness), in healthy male subjects. Br J Clin Pharmacol 87:3949–3960. https://doi.org/10.1111/bcp.14816

Gobbi F, Bottieau E, Bouchaud O et al (2018) Comparison of different drug regimens for the treatment of loiasis-A TropNet retrospective study. PLoS Negl Trop Dis 12:e0006917. https://doi.org/10.1371/journal.pntd.0006917

Gobbi F, Buonfrate D, Tamarozzi F et al (2019) Efficacy of high-dose albendazole with ivermectin for treating imported loiasis, Italy. Emerging Infect Dis 25:1574–1576. https://doi.org/10.3201/eid2508.190011

Grobusch MP, Kombila M, Autenrieth I et al (2003) No evidence of Wolbachia endosymbiosis with Loa loa and Mansonella perstans. Parasitol Res 90:405–408. https://doi.org/10.1007/s00436-003-0872-z

Gyapong JO, Chinbuah MA, Gyapong M (2003) Inadvertent exposure of pregnant women to ivermectin and albendazole during mass drug administration for lymphatic filariasis. Trop Med Int Health 8:1093–1101. https://doi.org/10.1046/j.1360-2276.2003.01142.x

Haselow NJ, Akame J, Evini C, Akongo S (2003) Programmatic and communication issues in relation to serious adverse events following ivermectin treatment in areas co-endemic for onchocerciasis and loiasis. Filaria J 2(Suppl 1):S10. https://doi.org/10.1186/1475-2883-2-S1-S10

Hawking F, Laurie W (1949) Action of hetrazan on filariasis and onchocerciasis. Lancet 2:146. https://doi.org/10.1016/s0140-6736(49)90993-9

Hawking F, Sewell P, Thurston JP (1950) The mode of action of hetrazan on filarial worms. Br J Pharmacol Chemother 5:217–238. https://doi.org/10.1111/j.1476-5381.1950.tb01010.x

Hemilembolo MC, Niama AC, Campillo JT et al (2023) Excess mortality associated with loiasis: confirmation be a new retrospective cohort study conducted in the Republic of Congo. Open Forum Infect DIs 10(3):ofad103. https://doi.org/10.1093/ofid/ofad103. eCollection 2023 Mar

Herrick JA, Legrand F, Gounoue R et al (2017) Posttreatment reactions after single-dose diethylcarbamazine or ivermectin in subjects with Loa loa infection. Clin Infect Dis 64:1017–1025. https://doi.org/10.1093/cid/cix016

Herrick JA, Makiya MA, Holland-Thomas N et al (2021) Infection-associated immune perturbations resolve 1 year following treatment for Loa loa. Clin Infect Dis 72:789–796. https://doi.org/10.1093/cid/ciaa137

Herrick JA, Metenou S, Makiya MA et al (2015) Eosinophil-associated processes underlie differences in clinical presentation of loiasis between temporary residents and those indigenous to Loa-endemic areas. Clin Infect Dis 60:55–63. https://doi.org/10.1093/cid/ciu723

Hewitt RI, White E (1947) Experimental chemotherapy of filariasis; effect of piperazine derivatives against naturally acquired filarial infections in cotton rats and dogs. J Lab Clin Med 32:1304–1313

Higazi TB, Klion AD, Boussinesq M, Unnasch TR (2004) Genetic heterogeneity in Loa loa parasites from southern Cameroon: a preliminary study. Filaria J 3:4. https://doi.org/10.1186/1475-2883-3-4

Hoang R, Temple C, Correia MS et al (2022) Characteristics of ivermectin toxicity in patients taking veterinary and human formulations for the prevention and treatment of COVID-19. Clin Toxicol (Phila) 60:1350–1355. https://doi.org/10.1080/15563650.2022.2134788

Hoegaerden MV, Ivanoff B, Flocard F et al (1987) The use of mebendazole in the treatment of filariases due to *Loa loa* and *Mansonella perstans*. Ann Trop Med Parasitol 81:275–282. https://doi.org/10.1080/00034983.1987.11812120

Hübner MP, Townson S, Gokool S et al (2021) Evaluation of the in vitro susceptibility of various filarial nematodes to emodepside. Int J Parasitol Drugs Drug Resist 17:27–35. https://doi.org/10.1016/j.ijpddr.2021.07.005

Kamgno J, Boussinesq M (2002) Effect of a single dose (600 mg) of albendazole on Loa loa microfilaraemia. Parasite 9:59–63. https://doi.org/10.1051/parasite/200209159

Kamgno J, Gardon J, Boussinesq M (2000) Analysis of the prevention of post-ivermectin Loa loa encephalopathy by administration of initial low dose. Med Trop (Mars) 60:275–277

Kamgno J, Nguipdop-Djomo P, Gounoue R et al (2016) Effect of two or six doses 800 mg of albendazole every two months on Loa loa microfilaraemia: a double blind, randomized, placebo-controlled trial. PLoS Negl Trop Dis 10:e0004492. https://doi.org/10.1371/journal.pntd.0004492

Kamgno J, Pion SD, Chesnais CB et al (2017) A test-and-not-treat strategy for onchocerciasis in Loa loa-endemic areas. N Engl J Med 377:2044–2052. https://doi.org/10.1056/NEJMoa1705026

Kamgno J, Pion SD, Mackenzie CD et al (2009) Loa loa microfilarial periodicity in ivermectin-treated patients: comparison between those developing and those free of serious adverse events. Am J Trop Med Hyg 81:1056–1061. https://doi.org/10.4269/ajtmh.2009.09-0356

Kamgno J, Pion SDS, Tejiokem MC et al (2007) Randomized, controlled, double-blind trial with ivermectin on Loa loa microfilaraemia: efficacy of a low dose (approximately 25 microg/kg) versus current standard dose (150 microg/kg). Trans R Soc Trop Med Hyg 101:777–785. https://doi.org/10.1016/j.trstmh.2007.03.018

Kanesa-thasan N, Douglas JG, Kazura JW (1991) Diethylcarbamazine inhibits endothelial and microfilarial prostanoid metabolism in vitro. Mol Biochem Parasitol 49:11–19. https://doi.org/10.1016/0166-6851(91)90125-p

Kashyap SS, Verma S, Voronin D et al (2019) Emodepside has sex-dependent immobilizing effects on adult Brugia malayi due to a differentially spliced binding pocket in the RCK1 region of the SLO-1 K channel. PLoS Pathog 15:e1008041. https://doi.org/10.1371/journal.ppat.1008041

Kivits M (1952) Four cases of fatal encephalitis with invasion of the cerebrospinal fluid by Microfilaria loa. Ann Soc Belg Med Trop (1920) 32:235–242

Klion AD, Eisenstein EM, Smirniotopoulos TT et al (1992) Pulmonary involvement in loiasis. Am Rev Respir Dis 145:961–963. https://doi.org/10.1164/ajrccm/145.4_Pt_1.961

Klion AD, Horton J, Nutman TB (1999) Albendazole therapy for loiasis refractory to diethylcarbamazine treatment. Clin Infect Dis 29:680–682. https://doi.org/10.1086/598654

Klion AD, Massougbodji A, Horton J et al (1993) Albendazole in human loiasis: results of a double-blind, placebo-controlled trial. J Infect Dis 168:202–206. https://doi.org/10.1093/infdis/168.1.202

Klion AD, Ottesen EA, Nutman TB (1994) Effectiveness of diethylcarbamazine in treating loiasis acquired by expatriate visitors to endemic regions: long-term follow-up. J Infect Dis 169:604–610. https://doi.org/10.1093/infdis/169.3.604

Kombila M, Duong TH, Ferrer A et al (1998) Short- and long-term action of multiple doses of ivermectin on loiasis microfilaremia. Am J Trop Med Hyg 58:458–460. https://doi.org/10.4269/ajtmh.1998.58.458

Kouam MK, Tchatchueng-Mbougua JB, Demanou M et al (2013) Impact of repeated ivermectin treatments against onchocerciasis on the transmission of loiasis: an entomologic evaluation in central Cameroon. Parasit Vectors 6:283. https://doi.org/10.1186/1756-3305-6-283

Krücken J, Harder A, Jeschke P et al (2012) Anthelmintic cyclcooctadepsipeptides: complex in structure and mode of action. Trends Parasitol 28:385–394. https://doi.org/10.1016/j.pt.2012.06.005

Lachau-Durand S, Lammens L, van der Leede B-J et al (2019) Preclinical toxicity and pharmacokinetics of a new orally bioavailable flubendazole formulation and the impact for clinical trials

and risk/benefit to patients. PLoS Negl Trop Dis 13:e0007026. https://doi.org/10.1371/journal.pntd.0007026
Lee KC, Ladizinski B, Federman DG (2012) Complications associated with use of levamisole-contaminated cocaine: an emerging public health challenge. Mayo Clin Proc 87:581–586. https://doi.org/10.1016/j.mayocp.2012.03.010
Legrand F, Herrick J, Makiya M et al (2021) A randomized, placebo-controlled, double-blind pilot study of single-dose humanized anti-IL5 antibody (reslizumab) for the reduction of eosinophilia following diethylcarbamazine treatment of Loa loa infection. Clin Infect Dis 73:e1624–e1631. https://doi.org/10.1093/cid/ciaa1365
Liu LX, Serhan CN, Weller PF (1990) Intravascular filarial parasites elaborate cyclooxygenase-derived eicosanoids. J Exp Med 172:993–996. https://doi.org/10.1084/jem.172.3.993
Loghry HJ, Yuan W, Zamanian M et al (2020) Ivermectin inhibits extracellular vesicle secretion from parasitic nematodes. J Extracell Vesicles 10:e12036. https://doi.org/10.1002/jev2.12036
Maizels RM, Denham DA (1992) Diethylcarbamazine (DEC): immunopharmacological interactions of an anti-filarial drug. Parasitology 105(Suppl):S49–S60. https://doi.org/10.1017/S0031182000075351
Makenga Bof J-C, Muteba D, Mansiangi P et al (2019) Analysis of severe adverse effects following community-based ivermectin treatment in the Democratic Republic of Congo. BMC Pharmacol Toxicol 20:49. https://doi.org/10.1186/s40360-019-0327-5
Martin-Prevel Y, Cosnefroy JY, Tshipamba P et al (1993) Tolerance and efficacy of single high-dose ivermectin for the treatment of loiasis. Am J Trop Med Hyg 48:186–192. https://doi.org/10.4269/ajtmh.1993.48.186
McGarry HF, Pfarr K, Egerton G et al (2003) Evidence against Wolbachia symbiosis in Loa loa. Filaria J 2:9. https://doi.org/10.1186/1475-2883-2-9
McGarry HF, Plant LD, Taylor MJ (2005) Diethylcarbamazine activity against Brugia malayi microfilariae is dependent on inducible nitric-oxide synthase and the cyclooxygenase pathway. Filaria J 4:4. https://doi.org/10.1186/1475-2883-4-4
Mengome L-E, Akue JP, Souza A et al (2010) In vitro activities of plant extracts on human Loa loa isolates and cytotoxicity for eukaryotic cells. Parasitol Res 107:643–650. https://doi.org/10.1007/s00436-010-1910-2
Metais A, Michalak S, Rousseau A (2021) Albendazole-related Loa loa encephalopathy. IDCases 23:e01033. https://doi.org/10.1016/j.idcr.2020.e1033. eCOllection 2021
Moreno Y, Nabhan JF, Solomon J et al (2010) Ivermectin disrupts the function of the excretory-secretory apparatus in microfilariae of Brugia malayi. Proc Natl Acad Sci USA 107:20120–20125. https://doi.org/10.1073/pnas.1011983107
Mrimi EC, Welsche S, Ali SM et al (2023) Emodepside for Trichuris trichiura and hookworm infection. N Engl J Med 388:1863–1875. https://doi.org/10.1056/NEJMoa2212825
Murgatroyd F, Woodruff AW (1949) Loiasis treated with hetrazan. Lancet 2:147–149. https://doi.org/10.1016/s0140-6736(49)90994-0
Muylle L, Taelman H, Moldenhauer R et al (1983) Usefulness of apheresis to extract microfilarias in management of loiasis. Br Med J (Clin Res Ed) 287:519–520
Nana-Djeunga HC, Lenou-Nanga CG, Donfo-Azafack C et al (2020) Collateral impact of community-directed treatment with ivermectin (CDTI) for onchocerciasis on parasitological indicators of Loa loa infection. Pathogens 9. https://doi.org/10.3390/pathogens9121043
Negesse Y, Lanoie LO, Neafie RC, Connor DH (1985) Loiasis: "Calabar" swellings and involvement of deep organs. Am J Trop Med Hyg 34:537–546. https://doi.org/10.4269/ajtmh.1985.34.537
Ngu JL, Chatelanat F, Leke R et al (1985) Nephropathy in Cameroon: evidence for filarial derived immune-complex pathogenesis in some cases. Clin Nephrol 24:128–134
Ngwewondo A, Wang M, Manfo FPT et al (2018) Filaricidal properties of Lantana camara and Tamarindus indica extracts, and Lantadene A from L. camara against Onchocerca ochengi and Loa loa. PLoS Negl Trop Dis 12:e0006565. https://doi.org/10.1371/journal.pntd.0006565

Nicolas P, Maia MF, Bassat Q et al (2020) Safety of oral ivermectin during pregnancy: a systematic review and meta-analysis. Lancet Glob Health 8:e92–e100. https://doi.org/10.1016/S2214-109X(19)30453-X

Njouendou AJ, Fombad FF, O'Neill M et al (2018) Heterogeneity in the in vitro susceptibility of Loa loa microfilariae to drugs commonly used in parasitological infections. Parasit Vectors 11: 223. https://doi.org/10.1186/s13071-018-2799-3

Norgan AP, Juskewitch JE, Pritt BS, Winters JL (2018) The use of cytapheresis in the treatment of infectious diseases. J Clin Apher 33:529–537. https://doi.org/10.1002/jca.21620

Nutman TB, Miller KD, Mulligan M, Ottesen EA (1986) Loa loa infection in temporary residents of endemic regions: recognition of a hyperresponsive syndrome with characteristic clinical manifestations. J Infect Dis 154:10–18. https://doi.org/10.1093/infdis/154.1.10

Nutman TB, Miller KD, Mulligan M et al (1988) Diethylcarbamazine prophylaxis for human loiasis. Results of a double-blind study. N Engl J Med

O'Connell EM, Bennuru S, Steel C et al (2015) Targeting filarial Abl-like kinases: orally available, food and drug administration-approved tyrosine kinase inhibitors are microfilaricidal and macrofilaricidal. J Infect Dis 212:684–693. https://doi.org/10.1093/infdis/jiv065

O'Connell EM, Nutman TB (2017) Reduction of Loa loa microfilaremia with imatinib - a case report. N Engl J Med 377:2095–2096. https://doi.org/10.1056/NEJMc1712990

O'Neill M, Njouendou JA, Dzimianski M et al (2018) Potential role for flubendazole in limiting filariasis transmission: observations of microfilarial sensitivity. Am J Trop Med Hyg 98:21–26. https://doi.org/10.4269/ajtmh.17-0390

Odedra A, Lalloo DG, Kennedy G et al (2019) Safety and effectiveness of apheresis in the treatment of infectious diseases: a systematic review. J Infect 79:513–520. https://doi.org/10.1016/j.jinf.2019.09.014

Oscanoa TJ, Amado J, Romero-Ortuno R, Carvajal A (2022) Hepatic disorders associated with the use of Ivermectin for SARS-CoV-2 infection in adults: a pharmacovigilance study in VigiBase. Gastroenterol Hepatol Bed Bench 15:426–429. https://doi.org/10.22037/ghfbb.v15i4.2383

Pakasa NM, Nseka NM, Nyimi LM (1997) Secondary collapsing glomerulopathy associated with Loa loa filariasis. Am J Kidney Dis 30:836–839. https://doi.org/10.1016/s0272-6386(97)90090-1

Pfarr KM, Krome AK, Al-Obaidi I et al (2023) The pipeline for drugs for control and elimination of neglected tropical diseases: 1. Anti-infective drugs for regulatory registration. Parasit Vectors 16:82. https://doi.org/10.1186/s13071-022-05581-4

Piessens WF, Beldekas M (1979) Diethylcarbamazine enhances antibody-mediated cellular adherence to Brugia malayi microfilariae. Nature 282:845–847. https://doi.org/10.1038/282845a0

Pionnier NP, Sjoberg H, Chunda VC et al (2019) Mouse models of Loa loa. Nat Commun 10:1429. https://doi.org/10.1038/s41467-019-09442-0

Pion SD, Nana-Djeunga H, Niamsi-Emalio Y et al (2020) Implications for annual retesting after a test-and-not-treat strategy for onchocerciasis elimination in areas co-endemic with Loa loa infection: an observational cohort study. Lancet Infect Dis 20:102–109. https://doi.org/10.1016/S1473-3099(19)30554-7

Pion SD, Tchatchueng-Mbougua JB, Chesnais CB et al (2019) Effect of a single standard dose (150-200 μg/kg) of ivermectin on Loa loa microfilaremia: systematic review and meta-analysis. Open Forum Infect Dis 6:ofz019. https://doi.org/10.1093/ofid/ofz019

Rajan TV, Shultz LD, Babu S et al (1998) Diethylcarbamazine (DEC) does not induce nitric oxide (NO) synthesis. Exp Parasitol 88:217–222. https://doi.org/10.1006/expr.1998.4247

Ramos-Zuniga R (2013) IMpact of antihelminthis treatment in neurosurgical patients with symptomatic neurocysticercosis. World Neurosurg 80(5):e141-2. https://doi.org/10.1016/j.wneu.2011.11.020. Epub 2011 Nov 30

Ranque S, Garcia A, Boussinesq M et al (1996) Decreased prevalence and intensity of Loa loa infection in a community treated with ivermectin every three months for two years. Trans R Soc Trop Med Hyg 90:429–430. https://doi.org/10.1016/s0035-9203(96)90535-0

Richard-Lenoble D, Kombila M, Burnier I, Maganga ML (1985) Filariasis in Gabon: treatment with mebendazole of filariasis due to M. perstans and Loa loa. Bull Soc Pathol Exot Filiales 78:485–491

Richard-Lenoble D, Kombila M, Rupp EA et al (1988) Ivermectin in loiasis and concomitant O. volvulus and M. perstans infections. Am J Trop Med Hyg 39:480–483. https://doi.org/10.4269/ajtmh.1988.39.480

Saito M, Armstrong M, Boadi S et al (2015) Clinical features of imported loiasis: a case series from the hospital for tropical diseases, London. Am J Trop Med Hyg 93:607–611. https://doi.org/10.4269/ajtmh.15-0214

Santiago-Stevenson D, Oliver-Gonzalez J, Hewitt R (1947) Treatment of filariasis bancrofti with 1-diethylcarbamyl-4-methylpiperazine hydrochloride (hetrazan). J Am Med Assoc 135:708–712. https://doi.org/10.1001/jama.1947.02890110026007

Santoni G, Morelli MB, Amantini C et al (2018) "Immuno-transient receptor potential ion channels": the role in monocyte- and macrophage-mediated inflammatory responses. Front Immunol 9:1273. https://doi.org/10.3389/fimmu.2018.01273

Schofield FD, Rowley RE (1961) The effect of prednisone on persistent microfilaremia during treatment with diethylcarbamazine. Am J Trop Med Hyg 10:849–854. https://doi.org/10.4269/ajtmh.1961.10.849

Shookhoff HB, Dwork KG (1949) Treatment of Loa loa infections with hetrazan. Am J Trop Med Hyg 29:589–593. https://doi.org/10.4269/ajtmh.1949.s1-29.589

Stanley SL, Kell O (1982) Ascending paralysis associated with diethylcarbamazine treatment of M. loa loa infection. Trop Doct 12:16–19. https://doi.org/10.1177/004947558201200107

Stingl P, Pierce PF, Connor DH et al (1988) Does dexamethasone suppress the Mazzotti reaction in patients with onchocerciasis? Acta Trop 45:77–85

Tabi T-E, Befidi-Mengue R, Nutman TB et al (2004) Human loiasis in a Cameroonian village: a double-blind, placebo-controlled, crossover clinical trial of a three-day albendazole regimen. Am J Trop Med Hyg 71:211–215. https://doi.org/10.4269/ajtmh.2004.71.211

Thompson JH (1956) ACTH as an adjunct to the treatment of loaiasis. Am J Trop Med Hyg 5:1103–1105. https://doi.org/10.4269/ajtmh.1956.5.1103

Tsague-Dongmo L, Kamgno J, Pion SDS et al (2002) Effects of a 3-day regimen of albendazole (800 mg daily) on Loa loa microfilaraemia. Ann Trop Med Parasitol 96:707–715. https://doi.org/10.1179/000349802125001933

Van Bogaert L, Dubois A, Janssens PG et al (1955) Encephalitis in loa-loa filariasis. J Neurol Neurosurg Psychiatr 18:103–119. https://doi.org/10.1136/jnnp.18.2.103

Van Hoegaerden M, Flocard F (1986) Treatment of recurrent, filarial, Calabar-type oedema with mebendazole. Trop Geogr Med 38:296–298

Van Hoegaerden M, Flocard F (1985) Mebendazole treatment of loaiasis. Lancet 1:1278. https://doi.org/10.1016/s0140-6736(85)92350-5

Veit O, Beck B, Steuerwald M, Hatz C (2006) First case of ivermectin-induced severe hepatitis. Trans R Soc Trop Med Hyg 100:795–797. https://doi.org/10.1016/j.trstmh.2006.02.003

Veletzky L, Eberhardt KA, Hergeth J et al (2022) Distinct loiasis infection states and associated clinical and hematological manifestations in patients from Gabon. PLoS Negl Trop Dis 16: e0010793. https://doi.org/10.1371/journal.pntd.0010793

Veletzky L, Hergeth J, Stelzl DR et al (2020) Burden of disease in Gabon caused by loiasis: a cross-sectional survey. Lancet Infect Dis 20:1339–1346. https://doi.org/10.1016/S1473-3099(20)30256-5

Verma S, Kashyap SS, Robertson AP, Martin RJ (2020) Diethylcarbamazine activates TRP channels including TRP-2 in filaria, Brugia malayi. Commun Biol 3:398. https://doi.org/10.1038/s42003-020-01128-4

Volpicelli L, De Angelis M, Morano A et al (2020) Encephalopathy in a patient with loiasis treated with albendazole: a case report. Parasitol Int 75:102006. https://doi.org/10.1016/j.parint.2019.102006

Wanji S, Chounna Ndongmo WP, Fombad FF et al (2018) Impact of repeated annual community directed treatment with ivermectin on loiasis parasitological indicators in Cameroon: implications for onchocerciasis and lymphatic filariasis elimination in areas co-endemic with Loa loa in Africa. PLoS Negl Trop Dis 12:e0006750. https://doi.org/10.1371/journal.pntd.0006750

Wanji S, Eyong E-E, Tendongfor N et al (2015) Parasitological, hematological and biochemical characteristics of a model of hyper-microfilariaemic loiasis (Loa loa) in the baboon (Papio anubis). PLoS Negl Trop Dis 9:e0004202. https://doi.org/10.1371/journal.pntd.0004202

Wanji S, Eyong E-EJ, Tendongfor N et al (2017) Ivermectin treatment of Loa loa hyper-microfilaraemic baboons (Papio anubis): assessment of microfilarial load reduction, haematological and biochemical parameters and histopathological changes following treatment. PLoS Negl Trop Dis 11:e0005576. https://doi.org/10.1371/journal.pntd.0005576

Whittaker C, Chesnais CB, Pion SDS et al (2022) Factors associated with variation in single-dose albendazole pharmacokinetics: a systematic review and modelling analysis. PLoS Negl Trop Dis 16:e0010497. https://doi.org/10.1371/journal.pntd.0010497

Wilson NE, Reaves BJ, Wolstenholme AJ (2021) Lack of detectable short-term effects of a single dose of ivermectin on the human immune system. Parasit Vectors 14:304. https://doi.org/10.1186/s13071-021-04810-6

Woodruff AW (1951) Destruction of microfilariae of Loa loa in the liver in loiasis treated with banocide (hetrazan). Trans R Soc Trop Med Hyg 44:470–480. https://doi.org/10.1016/s0035-9203(51)80025-7

Zahner H, Schares G (1993) Experimental chemotherapy of filariasis: comparative evaluation of the efficacy of filaricidal compounds in Mastomys coucha infected with Litomosoides carinii, Acanthocheilonema viteae, Brugia malayi and B. pahangi. Acta Trop 52:221–266. https://doi.org/10.1016/0001-706x(93)90010-9

Zofou D, Fombad FF, Gandjui NVT et al (2018) Evaluation of in vitro culture systems for the maintenance of microfilariae and infective larvae of Loa loa. Parasit Vectors 11:275. https://doi.org/10.1186/s13071-018-2852-2

Diagnosis of *Loa loa*: From Blood Identification to Innovative Approaches

Hugues C. Nana Djeunga, Jean-Paul Akue, Arnauld Efon Ekangouo, Linda Djune Yemeli, and Joseph Kamgno

Abstract

Loiasis is a parasitic disease due to the filarial nematode *Loa loa* and transmitted to humans via the bites of a tabanid fly of the genus *Chrysops*. The disease manifests itself in two main signs, the adult worm migration across the subconjunctiva ("eye worm") and evanescent localized subcutaneous swellings, called "Calabar Swellings." Serious adverse events (SAEs) can occur as a result of diethylcarbamazine or ivermectin treatment in people harboring a very high *L. loa* microfilarial load. Loiasis, has been therefore of primary significance in areas where it is coendemic to lymphatic filariasis or onchocerciasis. These last years, significant clinical impacts have been demonstrated. The diagnosis of this filarial infection has significantly evolved, from direct observation of parasites using microscopic-based examination techniques to various diagnostic approaches based on the detection of the parasite-related biomarkers. Molecular-based and flow cytometry techniques despite their high sensitivity, require expensive equipment and well-equipped laboratories, whereas microscopy and immunological-based techniques are costless and friendly user but less sensitive. Performance and cost-effectiveness vary greatly with regard to the

H. C. Nana Djeunga · A. Efon Ekangouo · L. Djune Yemeli
Higher Institute for Scientific and Medical Research (ISM), Yaoundé, Cameroon

J.-P. Akue
Department of Parasitology, Centre International of Medical Research of Franceville (CIRMF), Franceville, Gabon

J. Kamgno (✉)
Higher Institute for Scientific and Medical Research (ISM), Yaoundé, Cameroon

Department of Public Health, Faculty of Medicine and Biomedical Sciences, University of Yaoundé I, Yaoundé, Cameroon
e-mail: kamgno@crfilmt.org

J. P. Akue (ed.), *Loa loa: Latest Advances in Loiasis Research*,
https://doi.org/10.1007/978-3-031-49450-5_6

considered techniques, and the choice of the suitable diagnostic technique for loiasis depends on the application.

Keywords

Loiasis · Direct · Indirect · Innovative diagnostic tools · LoaScope · Leucoconcentration · PCR · RTPCR · RLFP · LAMP · ELISA · LFA

1 Introduction

During the last century, the diagnosis of loiasis was essentially based on the observation of the migration of adult worms under the conjunctiva of the eye, and the identification of microfilariae in the peripheral blood. It appeared that although these methods were specific, they were not sensitive enough to pick up most of the cases of loiasis. Indeed, it was shown that more than 70% of infected people do not harbor microfilariae in their peripheral blood, despite the presence of clinical signs related to the *Loa loa* infection. Since then, many approaches have been used to develop more sensitive and innovative techniques, ranging from direct detection of microfilariae to indirect diagnostic tests using molecular (genomic and circulating cell-free DNA), and immunological (IgG4-based test) biomarkers. This chapter focuses on the development of these diagnostic approaches, with a specific mention of the significance of the results yielded by these different tools.

2 Direct Diagnostic Tests

2.1 Examination of Blood

2.1.1 Fresh Blood

This diagnostic approach consists in the direct examination of 10–50 μL of a fresh blood sample using bright-field microscopy at low magnification (×100) (WHO 1991). *Loa loa* microfilariae can be identified by their size (230–300 μm), as well as their specific wriggling movements in the blood (Fig. 1). This technique is low cost, easy to implement, and gives a good indication of the presence of the parasite in the blood sample. However, sample should be collected on the daytime (10 AM to

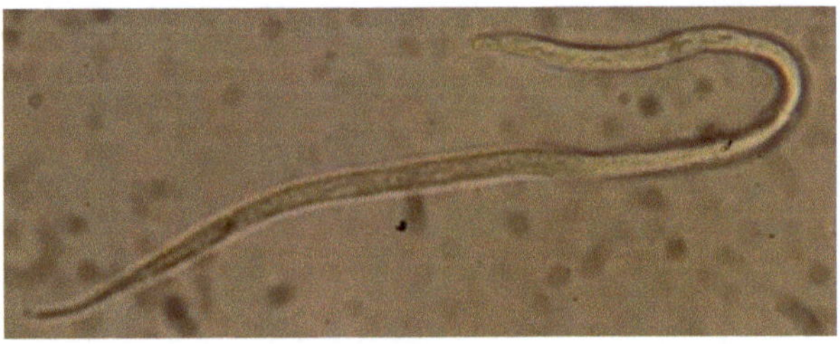

Fig. 1 *Loa loa* microfilaria in fresh blood (©ISM)

2 PM) according to the diurnal periodicity loiasis (Kamgno et al. 2009) and experienced technician is required for the differential diagnosis of *Wuchereria bancrofti* and *Mansonella perstans* which geographic distribution can overlap that of loiasis. Moreover, the standard microscopic examination of fresh blood does not allow to quantify the microfilarial load.

2.1.2 Calibrated Thick Blood Smear

The calibrated thick blood smear (CTBS) requires 50 μL of blood collected from finger prick in a nonheparinized capillary tube, spread on a slide and dried at room temperature. The dried preparation is dehemoglobinized (removal of hemoglobin) and stained with Giemsa. The preparation can be examined under a light microscope at high magnification (×400) for identification of *Loa loa* microfilariae (Fig. 2). The identification relies on the presence of the sheath, the distribution of nuclei and the sheath (Orihel et al. 1997). To differentiate between *Loa loa* and *Wucherera bacrofti*, the latter filaria measures 240–300/7.5–10 μm, has a tapered end without nuclei, its nuclei are well individualized ending before the tail, and its sheath is rarely colored by Giemsa. As for the *Loa loa* microfilaria, it measures 230–250/5–7 μm, with a tapered, slightly rounded tail and having nuclei to till the tip. The *Loa loa* microfilaria nuclei are nonindividualized and arranged to the tip of the tail, and its sheath is not stained with GiemAfter identification, the enumeration can be performed at low magnification (×100), and parasitemia expressed as the number of microfilariae per milliliter (mf/mL) of blood (WHO 1991). This technique is relatively inexpensive and routinely used in endemic areas to diagnose loiasis. Unlike direct diagnosis on fresh blood, CTBS enables a reliable estimate of the microfilarial load needed to identify individuals at risk of serious adverse events (SAEs) if treated with ivermectin. However, reliable microfilariae counting using the CTBS technique can become extremely tedious and approximate when the microfilarial load is very high. It is therefore difficult to use this technique for mass assessments in highly endemic areas. On the other hand, the sensitivity of the technique decreases with the parasitic load and its use is unreliable in persons with low numbers of circulating microfilariae (Noireau and Apembet 1990; Mischlinger

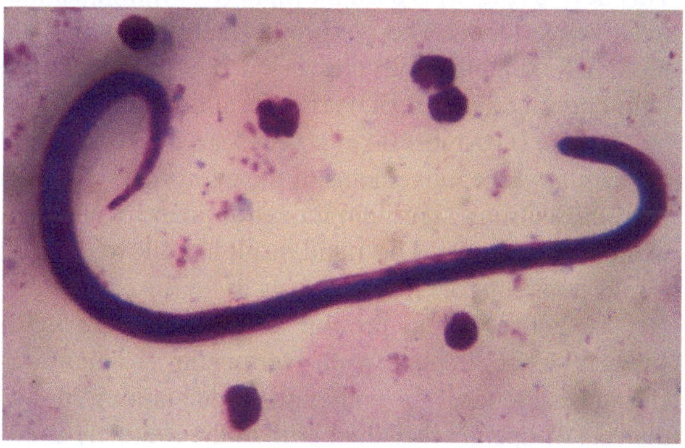

Fig. 2 With a Giemsa-stained *Loa loa* microfilaria (©ISM)

et al. 2021). This situation usually occurs in low-transmission areas, in areas under preventive chemotherapy or in travelers who have been transiently exposed to the infection.

2.2 Leucoconcentration

This technique is more sensitive than the fresh blood and calibrated thick blood smear as it uses a larger volume of blood. When the blood smear examination is negative, a concentration technique, such as the modified Knott's concentration technique can be routinely used as an alternative as well as filtration technique (Akue et al. 2011). The entire pellet retrieved is examined using a bright-field microscope at low magnification (×100) and microfilariae are identified and counted. The identification of *Loa loa* microfilariae relies on their size (230–250 μm) and the presence of a sheath. Although more sensitive than blood smears, this technique requires some equipment such as a centrifuge, and does not enable to quantify the microfilarial load.

2.3 Flow Cytometry

Flow cytometry has emerged as a powerful tool for counting and for classifying cells. The diagnosis of loiasis by flow cytometry uses a specific fluorescence pattern for microfilariae counts as previously described (Mouri et al. 2019). The main advantage of this technique is the counting of microfilaria. However, studies are yet to be conducted to evaluate its performance for the diagnosis of loiasis. Furthermore, due to the high cost and complexity of the equipment, its consideration for routine use is limited in low-income countries.

2.4 LoaScope

This device (Fig. 3) is used for rapid (2–3 min) quantitation of *L. loa* microfilariae in fresh whole blood. The LoaScope (first and second generation) uses the camera of the iPhone coupled with a reversed iPhone camera lens module for imaging and a light-emitting diode (LED) array for illumination to form a simple bright-field video microscope (D'Ambrosio et al. 2015). After peripheral blood collection (50 μL) in a flat capillary, the movements of the microfilariae are detected in seven fields using a machine learning algorithm trained to identify and enumerate *Loa loa* microfilariae in the blood sample. This rapid result has allowed the development of a new strategy for the treatment of onchocerciasis in loiasis coendemic areas, the so-called Test and Not Treat (TaNT), to prevent SAEs that occur in some loiasis heavily infected subjects when treated with ivermectin (Kamgno et al. 2017). This approach eliminates the need for preparation and staining of the blood samples, even though the cost of operations is higher than the routinely used CTBS.

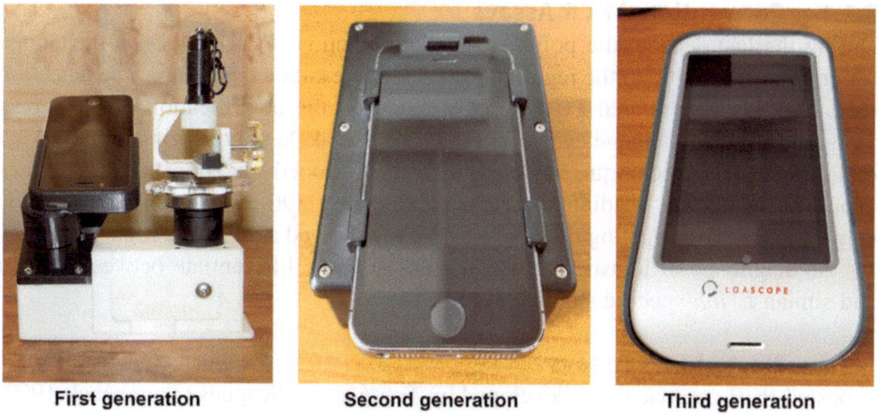

Fig. 3 Different generations of LoaScopes (©ISM)

2.5 Significance of Direct Diagnostic Tests

The direct diagnostic tests aim at detecting *Loa loa* microfilariae in the peripheral blood sample, which is direct evidence of the parasite. However, the enumeration of microfilariae on smear by microscopy is a nonautomated method, which may lack reproducibility and requires a trained technician. Although this limitation was overcome by the development of the LoaScope using machine learning, it should be noted that only about 30% of infected people harbor microfilariae in their peripheral blood, 70% being amicrofilaremic and cannot be identified by these direct diagnostic tools (Nanduri and Kazura 1989; Knopp et al. 2012).

3 Indirect Tests

New diagnostic approaches have been developed to overcome the shortcomings of direct diagnostic tests. These approaches include molecular and serological tests.

3.1 Molecular Diagnosis of Loiasis: PCR-Based Assays

Extraction and/or purification of nucleic acids should be performed as a prerequisite of every polymerase chain reaction (PCR)-based assay. This process can be carried out from different types of biological samples, including but not limited to total blood samples, buffy coats, and dried blood spots. Different techniques have been developed to extract and/or purify nucleic acids from these samples, including Chelex, phenol-chloroform, cetyltrimethylammonium bromide (CTAB), ion exchange chromatography using mini columns, and more recently magnetic beads-based methods (Iovieno et al. 2011; Bordelon et al. 2013; Wang et al. 2019).

3.1.1 Conventional PCR Assays

During the last decade, the polymerase chain reaction (PCR) has been developed using primers specific to the region of interest in the targeted genome. The detection of *L. loa* by conventional PCR initially relied on the amplification of the repeat 3 region (15r3) of the gene encoding for *Loa loa* 15-kDa polyprotein antigen (Ajuh et al. 1995). This technique has been further developed by Touré and colleagues (Touré et al. 1997), leading to the detection of a 396-bp DNA fragment. The conventional PCR targeting the *L. loa* 15r3 gene showed a relatively high sensitivity for the diagnostic of loiasis, but was not designed to differentiate between human and simian *L. loa*-specific strains.

3.1.2 Nested-PCR Assays

A nested-PCR approach was developed by Touré and colleagues (Touré et al. 1999) to enable differential diagnosis of simian *L. loa*. This technique is based on primers designed from the sequences of the repeat 3 region (15r3) of the gene encoding a *Loa loa* 15-kDa as previously described (Touré et al. 1998). The initial conventional PCR product (396 bp) is used as DNA template for a subsequent PCR using the nested primers set (the primer pair 151-3, and 15r3). The nested PCR products are run on gel electrophoresis for the detection of the DNA amplicons of interest. Positive samples are made up of 366 bp fragments specific to simian *L. loa*.

3.1.3 Real-Time PCR Assays

This technique which combines amplification and detection into a single step is based on the collection of data throughout the PCR process as it occurs. Two main markers were successfully used to diagnose loiasis using the real-time PCR, the first targeted by a set of primers and probe from the 15 kDa of *Loa loa* (Dieki et al. 2022a, b) and the second marker (LLMF72) described by Fink et al. (2011). In addition to internal transcribed spacer one (ITS1), a cell-free DNA marker (LL2643) was recently described by Bennuru et al. (2023) for the diagnosis of loiasis using real-time PCR from blood and urine samples.

3.1.4 Restriction Fragment Length Polymorphism

Ribosomal deoxyribonucleic acid (rDNA) and internal transcribed spacer region (ITS1) were also targeted for differential diagnosis of filarial infections, including loiasis, by using a species-specific polymerase chain reaction-restriction fragment length polymorphism (PCR-RFLP) protocol developed by Jiménez et al. (2011). Indeed, subsequent to the conventional PCR, the ITS1-PCR product is digested with the restriction endonuclease to generate species-specific fragments sizes, which enables differential identification of three filarial species namely *Loa loa*, *Mansonella perstans,* and *Wuchereria bancrofti* (Jiménez et al. 2011). Although the RFLP presents a high sensitivity and reproducibility, the technique is highly expensive and requires laborious and technically demanding methodological procedures.

3.1.5 Loop-Mediated Isothermal Amplification

Loop-mediated isothermal amplification (LAMP) assay has emerged as a new molecular-based diagnostic approach with the potential to replace the PCR (Abdul-Ghani et al. 2012). Different LAMP approaches have been developed for the diagnosis of loiasis, the first one using a 839 bp *Loa loa*-specific repetitive DNA sequence (GenBank accession no.: M34259.1) (Klion et al. 1991; Fernandez-Soto et al. 2014; Febrer-Sendra et al. 2022), and the other using the RF4 LAMP assay (Poole et al. 2015). As major equipment are portable and user-friendly, the LAMP is a good alternative for use in the field. Additionally, LAMP showed similar performances in terms of sensitivity and specificity compared to the other diagnostic methods for loiasis such as microscopy, PCR, and Nested PCR (Ta-Tang et al. 2022). However, the main disadvantages are the high price of reagents and the inability to detect more than one filarial species simultaneously (Ta-Tang et al. 2022).

3.1.6 Significance of *L. loa* DNA Detection in a Sample

All the PCR-based assays aim at detecting *Loa loa* parasite DNA in the sample. The presence of DNA means the presence of alive parasite or an active infection state. In some cases, it might happen that microfilaria or adult worm have not been documented in these samples, but the DNA is present, so-called cryptic infection.

3.2 Serological Diagnosis of Loiasis

3.2.1 Enzyme-Linked Immunosorbent Assays

This technique in general is used for the detection of antibody or antigen in a sample. Many formats have been developed using crude or homologous extract of the parasite. These homemade procedures (Akue et al. 1994; Dieki et al. 2022a, b) can be easily set up in many laboratories. Nowadays, other enzyme-linked immunosorbent assays (ELISA) formats are emerging and used at point of care such as the pan-filarial ELISA (Gobbi et al. 2020). Prior to performing ELISA *per se*, antigens can be prepared from different samples (including but not limited to plasma, serum, or dried blood spots), and the protein content can be measured according to the Bradford method.

3.2.2 Lateral Flow Assay for *Loa loa*

Many lateral flow assays (LFA) have been developed during the last decades, the most promising one being the Loa Antibody Rapid Test (LART), based on the detection of IgG against the 148 amino acid sequence of L1 SXP-1 (Pedram et al. 2017). Indeed, when an IgG antibody specific for Ll-SXP-1 is present, one of its Fab portions can bind to the reporter nanoparticle while the other can be captured by the test line, thus leading to an accumulation of nanoparticles at the test line, and therefore to a visual signal. This design can detect any antibody irrespectively of its isotype. This protein has 51–53% sequence identity with *Wuchereria bancrofti* and *Onchocerca volvulus*, with sensitivity of 94% and specificity of 82–88% versus other filariae. In addition, the test can be used without additional equipment.

However, the LART only detects exposition to infection, and cannot ascertain whether the transmission is active or not (past infection).

3.2.3 Identification of Antigens by Western Blot Test

To date, there is no standard western blot test for *Loa loa*. However, to confirm the presence of *Loa loa* parasite in some cases, identification of antigens by blot can be needed. A homemade procedure was recently developed and can be carried out in two steps, including (1) the separation of crude antigens by SDS-PAGE, and (2) transfer on nitrocellulose paper and blot with IgG4 coupled with enzyme (Dieki et al. 2022a, b).

3.2.4 Other Serological Assays

Circulating filarial antigens can be detected in sera of patients with *Loa loa* filarial infection by coelectrosyneresis (Co-ES) (Walker-Deemin et al. 1996). Also, circulating immune complexes associated with *Loa loa* antigen can be detected in sera of *Loa loa* infected individuals, either "occult" or microfilaremic infections using Polyethylene glycol enzyme-linked immunosorbent assay (PEG-ELISA) (Jaoko 1995). Attempts were also done on urine (Eyang-Assengone 2015).

3.2.5 Significance of Outcomes of Serological Assays

These tests based on antibodies detection, need a good interpretation. The presence of antibodies does not mean necessarily the presence of infectious agents. Antibodies can remain in body a long time after a past infection. Their presence means the individual has been in contact with *Loa loa*. It is therefore difficult to ascertain a current or past infection status based only on antibodies-based tests.

4 Conclusion

The diagnosis of loiasis has significantly evolved over time. Initially based on the direct identification of the parasite in blood samples (blood smear) and relying on simple and easy to perform methods, the procedures significantly evolved to more reliable molecular- and serological-based techniques capturing the signature of the parasite in different type of samples for performance (sensitivity and specificity) purpose.

References

Abdul-Ghani R, Al-Mekhlafi AM, Karanis P (2012) Loop-mediated isothermal amplification (LAMP) for malarial parasites of humans: would it come to clinical reality as a point-of-care test? Acta Trop 122(3):233–240

Ajuh PM, Akue JP, Boutin P, Everaere S, Egwang TG (1995). Loa loa: structural diversity of a 15-kDa repetitive antigen. Exp Parasitol 81(2):145–53. https://doi.org/10.1006/expr.1995.1103

Akue JP, Egwang TG, Devaney E (1994) High levels of parasite-specific IgG4 in the absence of microfilaremia in Loa loa infection. Trop Med Parasitol 45(3):246–248

Akue JP, Nkoghe D, Padilla C, Moussavou G, Moukana H, Mbou RA, Ollomo B, Leroy EM (2011) Epidemiology of concomitant infection due to Loa loa and Mansonella perstans in Gabon. PLoS Negl Trop Dis 5(10):e1329. https://doi.org/10.1371/journal.pntd.0001329

Bennuru S, Kodua F, Drame PM, Dahlstrom E, Nutman TB (2023) A novel, bioinformatically informed highly sensitive NAAT assay for the diagnosis of loiasis and its use for detection of circulating cell free DNA. J Infect Dis:jiad186. https://doi.org/10.1093/infdis/jiad186. Epub ahead of print

Bordelon H, Russ PK, Wright DW, Haselton FR (2013) A magnetic bead-based method for concentrating DNA from human urine for downstream detection. PLoS One 8(7):e68369. https://doi.org/10.1371/journal.pone.0068369

D'Ambrosio MV, Bakalar M, Bennuru S, Reber C, Skandarajah A, Nilsson L, Switz N, Kamgno J, Pion S, Boussinesq M, Nutman TB, Fletcher DA (2015) Point-of-care quantification of blood-borne filarial parasites with a mobile phone microscope. Sci Transl Med 7(286):286re4. https://doi.org/10.1126/scitranslmed.aaa3480

Dieki R, Eyang Assengone ER, Nsi Emvo E, Akue JP (2022a) Profile of loiasis infection through clinical and laboratory diagnostics: the importance of biomarkers. Trans R Soc Trop Med Hyg: trac116. https://doi.org/10.1093/trstmh/trac116. Epub ahead of print

Dieki R, Nsi-Emvo E, Akue JP (2022b) The human filaria *Loa loa*: update on diagnostics and immune response. Res Rep Trop Med 13:41–54. https://doi.org/10.2147/RRTM.S355104

Eyang-Assengone EAR (2015) Analyse des complexes immuns circulants et des protéines urinaires des personnes infectées par Loa loa à l'aide des sous classes d'immunoglobuline G1 et G4. [Analysis of circulating immune complexes and urinary proteins from Loa loa infected individuals using subclasses of immunoglobulin G1 and G4]. [Master's thesis]. Université de Ouagadougou, Burkina Faso

Febrer-Sendra B, Fernández-Soto P, Crego-Vicente B, Diego JG, Ta-Tang TH, Berzosa P, Nguema R, Ncogo P, Romay-Barja M, Herrador Z, Benito A, Muro A (2022) Colorimetric and real-time loop-mediated isothermal amplification (LAMP) for detection of *Loa loa* DNA in human blood samples. Diagnostics (Basel) 12(5):1079. https://doi.org/10.3390/diagnostics12051079

Fernández-Soto P, Mvoulouga PO, Akue JP, Abán JL, Santiago BV, Sánchez MC et al (2014) Development of a highly sensitive loop-mediated isothermal amplification (LAMP) method for the detection of *Loa loa*. PLoS One 9(4):e94664. https://doi.org/10.1371/journal.pone.0094664

Fink DL, Kamgno J, Nutman TB (2011) Rapid molecular assays for specific detection and quantitation of *Loa loa* microfilaremia. PLoS Negl Trop Dis 5(8):e1299. https://doi.org/10.1371/journal.pntd.0001299

Gobbi F, Buonfrate D, Boussinesq M, Chesnais CB, Pion SD, Silva R et al (2020) Performance of two serodiagnostic tests for loiasis in a Non-Endemic area. PLoS Negl Trop Dis 14(5): e0008187. https://doi.org/10.1371/journal.pntd.0008187

Iovieno A, Miller D, Lonnen J, Kilvington S, Alfonso EC (2011) Extraction of Acanthamoeba DNA by use of Chelex resin. J Clin Microbiol 49(1):476–477. https://doi.org/10.1128/JCM.01795-10. Epub 2010 Nov 17

Jaoko WG (1995) Loa loa antigen detection by ELISA: a new approach to diagnosis. East Afr Med J 72(3):176–179

Jiménez M, González LM, Carranza C, Bailo B, Pérez-Ayala A, Muro A, Pérez-Arellano JL, Gárate T (2011) Detection and discrimination of Loa loa, Mansonella perstans and Wuchereria bancrofti by PCR-RFLP and nested-PCR of ribosomal DNA ITS1 region. Exp Parasitol 127(1):282–286. https://doi.org/10.1016/j.exppara.2010.06.019. Epub 2010 Jun 19

Kamgno J, Pion SDS, Mackenzie CD, Thylefors B, Boussinesq M (2009) Loa loa microfilarial periodicity in ivermectin-treated patients: comparison between those developing and those free of serious adverse events. Am J Trop Med Hyg 81:1056–1061

Kamgno J, Pion SD, Chesnais CB, Bakalar MH, D'Ambrosio MV, Mackenzie CD, Nana-Djeunga HC, Gounoue-Kamkumo R, Njitchouang GR, Nwane P, Tchatchueng-Mbouga JB, Wanji S, Stolk WA, Fletcher DA, Klion AD, Nutman TB, Boussinesq M (2017) A test-and-not-treat strategy for onchocerciasis in Loa loa-endemic areas. N Engl J Med 377(21):2044–2052. https://doi.org/10.1056/NEJMoa1705026

Klion AD, Raghavan N, Brindley PJ, Nutman TB (1991) Cloning and characterization of a species-specific repetitive DNA sequence from *Loa loa*. Mol Biochem Parasitol 45:297–305
Knopp S, Steinmann P, Hatz C, Keiser J, Utzinger J (2012) Nematode infections: filariases. Infect Dis Clin N Am 26(2):359–381
Mischlinger J, Manego RZ, Mombo-Ngoma G, Ekoka Mbassi D, Hackbarth N, Ekoka Mbassi FA, Davi SD, Kreuzmair R, Veletzky L, Hergeth J, Ndoumba WN, Pitzinger P, Groger M, Matsiegui PB, Adegnika AA, Agnandji ST, Lell B, Ramharter M (2021) Diagnostic performance of capillary and venous blood samples in the detection of Loa loa and Mansonella perstans microfilaraemia using light microscopy. PLoS Negl Trop Dis 15(8):e0009623. https://doi.org/10.1371/journal.pntd.0009623
Mouri O, Ezzine N, Haddad E, Achouri L, Parizot C, Thellier M, Piarroux R (2019) New promising method to assess microfilarial Loa loa load on the peripheral blood. Diagn Microbiol Infect Dis 95(4):114887. https://doi.org/10.1016/j.diagmicrobio.2019.114887. Epub 2019 Aug 12
Nanduri J, Kazura JW (1989) Clinical and laboratory aspects of filariasis. Clin Microbiol Rev 2(1): 39–50
Noireau F, Apembet JD (1990) Comparison of thick blood smear and saponin haemolysis for the detection of Loa loa and Mansonella perstans infections. J Trop Med Hyg 93(6):390–392
Orihel TC, Ash R, Ramachandran CP (1997) Planches pour identification des filaires. Organisation Mondiale de la Santé (OMS), Geneva. ISBN 9242544892
Pedram B, Pasquetto V, Drame PM, Ji Y, Gonzalez-Moa MJ, Baldwin RK et al (2017) A novel rapid test for detecting antibody responses to *Loa loa* infections. PLoS Negl Trop Dis 11(7): e0005741. https://doi.org/10.1371/journal.pntd.0005741
Poole CB, Ettwiller L, Tanner NA, Evans TC Jr, Wanji S, Carlow CKS (2015) Genome filtering for new DNA biomarkers of Loa loa infection suitable for loop-mediated isothermal amplification. PLoS One 10(9):e0139286. https://doi.org/10.1371/journal.pone.0139286
Ta-Tang TH, Berzosa P, Rubio JM, Romay-Barja M, Ncogo P, Agudo D, Herrador Z, Cerrada-Gálvez L, Benito A (2022) Evaluation of LAMP for the diagnosis of Loa loa infection in dried blood spots compared to PCR-based assays and microscopy. Mem Inst Oswaldo Cruz 116: e210210. https://doi.org/10.1590/0074-02760210210
Touré FS, Bain O, Nerrienet E, Millet P, Wahl G, Toure Y, Doumbo O, Nicolas L, Georges AL, McReynolds LA, Egwang TG (1997) Detection of Loa loa-specific DNA in blood from occult-infected individuals. Exp Parasitol 86(3):163–170., ISSN 0014-4894,. https://doi.org/10.1006/expr.1997.4168
Touré FS, Kassambara L, Williams T, Millet P, Bain O, Georges AJ et al (1998) Human occult loiasis: improvement in diagnostic sensitivity by the use of a nested polymerase chain reaction. Am J Trop Med Hyg 59(1):144–149
Toure FS, Leroy EM, Mavoungou E, Egwang TG (1999) Sequence conservation of repeat 3 region of the gene coding for the 15 kDa polyprotein within human and simian Loa loa. J Med Primatol 28(2):57–61. https://doi.org/10.1111/j.1600-0684.1999.tb00251.x
Walker-Deemin A, Kombila M, Mouray H, Rondi ML, Nguiri C, Richard-Lenoble D (1996) Detection of circulating antigens in Gabonese patients with Loa loa filariasis. Trop Med Int Health 1(6):772–778. https://doi.org/10.1111/j.1365-3156.1996.tb00109.x
Wang YS, Dai TM, Tian H, Wan FH, Zhang GF (2019) Comparative analysis of eight DNA extraction methods for molecular research in mealybugs. PLoS One 14(12):e0226818. https://doi.org/10.1371/journal.pone.0226818
WHO (1991) Basic laboratory methods in medical parasitology. World Health Organization, Geneva, 69 pp. https://apps.who.int/iris/handle/10665/40793

Immune Mechanism in Loiasis and Interactions with Other Infections

Jean Paul Akue and Roland Dieki

Abstract

The immune response caused by loiasis in the infected host has been studied during the past few decades. Interesting features have been revealed and diverse reactions have been reported. These include the mechanisms of protective immunity, cross-linking, anergy, and immune regulation, but also the impact of coinfection with other pathogens. Further, antigens and targets of these immune reactions have been identified. This chapter will cover the immunological mechanisms described for loiasis to date, and reveal the current gaps in knowledge that remain to be filled in the future.

Keywords

Loa loa · Infective larva · Microfilaria · IgG · T-helper · Antigens · Anergy · Coinfection

1 Introduction

Loa loa infection is characterized by a striking feature: Approximately 30% of infected individuals are microfilaremic, and 70% are amicrofilaremic. This status suggests immune mechanisms are involved either in the maintenance of the microfilaremia state or in resistance through the state of amicrofilaremia. In fact, microfilaremia can be maintained through the mechanism of downregulation of the immune response, including parasite escape of the host immune response. At the same time, amicrofilaremia can be the result of resistance against infection through protective immunity induced by protective antigens or through concomitant

J. P. Akue (✉) · R. Dieki
Neglected Parasitosis, CIRMF, Franceville, Gabon

J. P. Akue (ed.), *Loa loa: Latest Advances in Loiasis Research*,
https://doi.org/10.1007/978-3-031-49450-5_7

immunity targeting stage-specific antigens. In loiasis, many of these mechanisms have been reported either in field studies or in experimental models. In this chapter, we discuss these mechanisms and how they have been studied in loiasis.

2 Protective Immunity in Loiasis

Resistance or protective immunity in loiasis is demonstrated by the large proportion of amicrofilaremic individuals in endemic areas. While a smaller proportion of individuals harbor detectable microfilariae, the term "amicrofilaremia" refers to infection with adult worms without the appearance of microfilariae. Both states, microfilaremic and amicrofilaremic, are temporally and spatially stable over long periods of time. For example, studies by Van Hoegarden et al. (1987) in Gabon have shown that residents of endemic villages, who were followed up for 1 year, maintained their status as amicrofilaremic or microfilaremic during the follow-up period without significant change. Subsequent studies in Cameroon that followed up a community for as long as 23 years presented similar results (Mogoung-Wafo et al. 2019) despite the high intensity of disease transmission. Further, these findings are backed up by observations in experimental models of *Mandrillus sphinx*. In this model, mandrills were inoculated against infective-stage larvae (L3) irradiated with 40 krad. The experiment was carried out as follows: one group of six mandrills were inoculated with three doses of 50 L3 per dose of normal *L. loa* and six other mandrills were inoculated with three doses of 50 irradiated (40 krad) L3 per dose. Both groups of mandrills were challenged with a single dose of 100 live normal L3. After the challenge, both groups were followed up for 2 years with regular blood examinations for the detection of microfilariae. The results showed that there was a delay in microfilaremia appearance in the group vaccinated with irradiated L3, and the peak microfilaremia in this group was lower compared to the group vaccinated with normal infective L3 (Akue et al. 2003). This relationship between infective L3 and microfilariae is interesting, and suggests an immunity that acts on the larvae rather than on the adult filarial, called "concomitant immunity." This hypothesis is further supported by a study that was carried out in endemic villages including permanent residents. The study showed that microfilaria prevalence may increase up to a certain age and then it decreases afterwards. Simultaneously, the specific immunoglobulin (Ig) G4 level was measured as this subclass suggests chronicity of the infection. IgG4 levels increase with age, reach a plateau at the age of approximately 45 years, and remain there until senescence (Djikeussi and Akue 2014). This finding suggests that older adults confirm the chronicity of the infection and stimulate the production of IgG4. In parallel, with increased IgG4, microfilaria production diminishes, without the production of new microfilariae. This decrease in microfilariae is due to nonestablishment of new worms due to the action of immunity on new L3 (Fig. 1; Djikeussi and Akue 2014). This observation is supported by studies of Balb/c mice, which demonstrate that primary infection with *L. loa* may induce immune cell memory. This immune cell memory can stimulate the

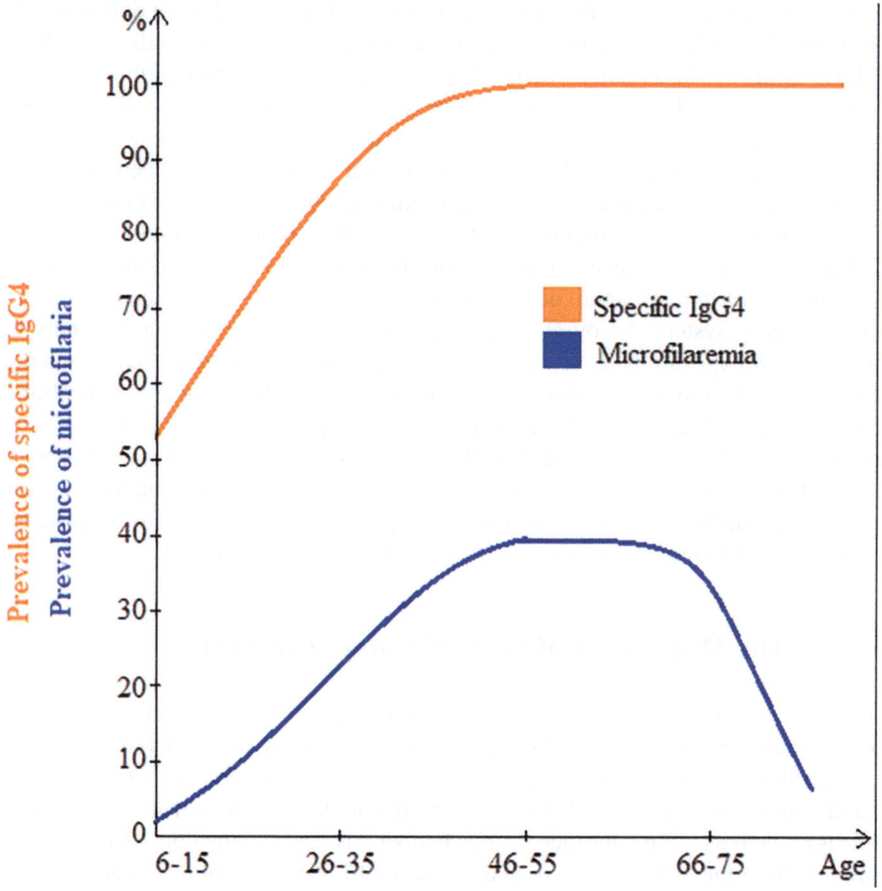

Fig. 1 The kinetics of IgG4 and microfilaremia in a population exposed to homogeneous transmission intensity (adapted from Djikeussi and Akue 2014). Specific IgG4 (orange curve) represents the presence of adult worm, which continuously increases up to the age of 46–55 years, then reaches a plateau from 55 to over 75 years of age, while the prevalence of microfilariae increases up to 46–55 years and then starts declining after 66 years of age, indicating the low productivity in older adults (blue curve)

production of cytokines, chemokines, and immunoglobulins necessary for protection against reinfection (Chunda et al. 2020).

3 The Mechanism of Anergy in *Loa loa*

A specific feature of loiasis is the appearance of microfilaremia in approximately 30% of infected individuals. Although host genetics has been suggested as a contributing factor, immune mechanisms certainly play an important role. Studies of humans in endemic areas have shown two different profiles of cytokine synthesis in microfilaremic and amicrofilaremic individuals. These profiles are dominated by Th1 (interleukin [IL]-2, interferon-gamma [IFN-gamma]) and Th2 cells (IL-4, IL-5) (Baize et al. 1997). A simultaneous downregulation of T-helper proliferation has also been reported among the microfilaremic group. Interestingly, the humoral response differed from the cellular response, as specific IgG4 was high in microfilaremic individuals. The mechanism that underlies this phenomenon is thought to be due to the lack of an appropriate costimulatory signal owing to an interaction between B7 (antigen-presenting cells [APC]) and CTLA-4 (T cells). This is suggested by the analysis of the cellular response in animal models, where the appearance of microfilariae downregulates cell proliferation by acting on the costimulatory system (Leroy et al. 1997). Another way to induce anergy may be the *in utero* exposure to *L. loa* antigens. It has been demonstrated that newborns from infected mothers were more susceptible to filarial infection than the newborns of uninfected mothers due to an expansion of regulatory T cells (Tregs). This expansion of Tregs diminished the Th1, Th2, and Th17 responses (Ateba-Ngoa et al. 2014). Further, the transmission of filariae between the mother and the newborn may additionally result in an increased level of total oxidative stress, possibly influencing the immunity of the newborn (Lemnui et al. 2020).

4 The Mechanism of Cross-Linking in *Loa loa*

The simultaneous elevation of specific IgG4 and IgE in loiasis infection has raised some questions about the underlying mechanism. As in other filarial infections, competition for the same antigenic site has been suggested. In loiasis, it has additionally been suggested that this may be due to the potential effect of a blocking of IgE activity. The proposed mechanism is an interaction between the immunoreceptor tyrosine-based activation motif (ITAM) and immunoreceptor tyrosine-based inhibitory motif (ITIM), which are fixed on cells including eosinophils, macrophages, neutrophil natural killer (NK) cells, and dendritic cells. Indeed, the activation of these receptors follows a linking of *L. loa* antigens, IgG4, and IgE. This leads to a series of phosphorylation and dephosphorylation using phosphatases containing the inositol phosphatase (SHP-1) and SHP-2, which results in two possible outcomes: (a) either the suppression of cell activation, which can block allergy, (b) or the activation of cells, which could result in allergic reactions (Dieki et al. 2022).

4.1 Regulation of Immune Response to *Loa loa*

Although specific studies on the regulation of immune response in loiasis have not been performed, different trends in the profiles of loiasis have been reported. This is characterized by the appearance of Th1 or Th2 cytokines, while an increase or decrease in some immunoglobulin isotypes, such as IgG1 and IgG4, has also been reported. This suggests a mechanism of regulation underlying these effects. Studies on mothers and the cord blood of their offspring (Ateba-Ngoa et al. 2014) showed that regulatory cells exist in both the mother's blood and the blood from the cord of the newborn. It is known that $CD4^{+}CD25^{+}$hiFOX3P3 are potent Tregs that exert a suppressive effect on both the proliferation and production of cytokines, and that $CD4^{+}$ $Tbet^{+}$ are promotors of antitumoral activity by improving the cytotoxicity of $CD4^{+}$ ROR_{γ} T^{+}, which in turn are promotors of transcription factors of Th17 cells. The markers FOX3, bet, and ROR^{γ}, were used to measure the distribution of these cells in offspring and their mother. This study showed an inverse correlation between $CD4^{+}CD25^{+}$hi FOX3P3 and both bet and ROR, thus confirming the existence of such cells in both maternal and neonatal cord blood. The study supported the existence of a regulatory mechanism, similar to those shown in other filarial infections in which Tregs play an active role in filarial clinical expression. It is known that IL-10 and IgG4 are produced at a high concentration in *L. loa* infection. IL-10 and Tregs promote the production of IgG4 by B cells (Nutman et al. 1986; Hussein et al. 1992) and, further, the induction of IgG4 by Tregs $CD4^{+}$ $CD25^{+}$ $FOXP3^{+}$ has been reported. Microfilaremic loiasis induces tolerogenic APC, which in turn produce IL-10, TGF-beta, Tregs, $CD4^{+}$ $CD25^{+}$ $FOXP3^{+}$, and regulatory type 1 (Tr1) cells, which interact with B cells to enhance the production of IgG4. By contrast, in amicrofilaremic individuals, the production of IL-12, IFN-gamma, and IL-4, which is characteristic of Th1, Th2, and Th17, was increased, probably under the action of regulatory cells. These cytokines act on B cells leading to the production of IgG1, IgG2, IgG3, and IgE and thereby they induce effector mechanisms responsible for the killing of microfilariae or infective L3. The regulatory role of IL-10, IL-19, and IL-23 was recently shown in the immune modulation of *L. loa* infection (Riccardi and Nutman 2021).

4.2 Mechanisms of Immune Evasion by *L. loa*

Loa loa employs different mechanisms to escape the host immune response. One of the mechanisms reported is the evasion of complement binding. It was shown in vivo that a regulatory protein, called factor H, is used by *L. loa* microfilariae on their sheath to block the activation of the complement cascade (Haapsalo et al. 2009). Similarly, human albumin was shown on the microfilaria surface (Egwang et al. 1988b). Another mechanism used by *L. loa* is immune suppression or immune modulation in favor of parasite survival (Leroy et al. 1997). In the *L. loa* genome, a gene coding for an antagonist of IL-5 has been found, which inhibits macrophage migration, the presence of lectin type C, galactin, jacalin, and genes coding for lipopolysaccharide-binding molecules. Further, the presence of serpin and cystatin genes, which are known to interfere with antigen presentation to T cells, has been reported (Hartman and Lucius 2003). The appearance of microfilariae in the peripheral blood coincides with the suppression or downregulation of the immune response (Leroy et al. 1997). It seems that molecular mimicry may also be one of the mechanisms used by *L. loa* to escape the host defense system. Although there are no data to confirm the activities for members of IL-16, antagonist of IL-5 receptor, factors for the regulation of IFN-gamma, the homolog of the suppression signal of cytokines (SOCS-7), however, there were genes identified in *L. loa* genome.

4.3 Immune Response in Human Host Coinfected with Other Pathogens

The simultaneous presence of *L. loa* and other pathogens in the same individual is common in endemic countries. This situation involves many confounding factors and may lead to misdiagnosis of the infection. A coinfection of three parasites has been reported in Cameroon (Papa Drame et al. 2016): *L. loa, Mansonella perstans,* and *Plasmodium falciparum.* The cooccurrence of these three parasites seems to be associated with elevated microfilaremia of *L. loa*, while the other parasites did not show a significant influence. In the coinfection of microfilaremic *L. loa* and HIV-1, there was an elevation of IgE exclusively in microfilaremic HIV-1-infected individuals, in contrast to the decrease of IgG4 and IgG3 levels in the same individuals (Njambe Priso et al. 2018). Similarly, it was reported that in high-transmission zones of *L. loa,* the proliferation of peripheral blood mononuclear cells (PBMC) is low compared to that in zones of low transmission. This effect remained even when cells were stimulated with purified protein derivative (PPD) from mycobacterium, an antigen of bacterial origin (Akue and Devaney 2002). These observations demonstrate the multiple outcomes of immunity in the case of coinfection and the likely influence of *L. loa* or other pathogens in the same environment. *Loa loa* serves either as a trigger or as a cofactor in changes of the immune profile and the clinical outcome in the host. These aspects need to be studied further. However, in the case of coinfection between *L. loa* and HTLV-1, a mechanism leading to a negative outcome can be hypothesized (Fig. 2): HTLV-1 has a region in its genome that carries the *pX* gene. This gene produces a molecule called TAX, which is implicated in viral replication and transformation of CD4 T cells. When HTLV-1 penetrates CD4 T cells, it induces proliferation of CD4 T cells carrying the virus with the *pX* gene. This is followed by the synthesis of inflammatory cytokines IL-1, IL-5, and TNF-gamma. At the same time, the presence of high-density *L. loa* microfilaremia generates a Th2-type environment dominated by inflammatory cytokine production with IL-5 and IL-4. The high levels of these proinflammatory cytokines lead to an exacerbation of inflammatory responses, which may cause adult T cell leukemia lymphoma (ATLL) or encephalopathy (Dieki 2023).

4.4 *Loa loa* Antigens as a Target of the Host Immune Response

Three main sources of antigens can be found in parasites: surface antigens as well as somatic excreted and secreted antigens. In *L. loa,* several antigens have been identified by their molecular weight as well as their homolog to antigens of lymphatic filaria (Akue 1995). The antigens show a large variability in molecular size,

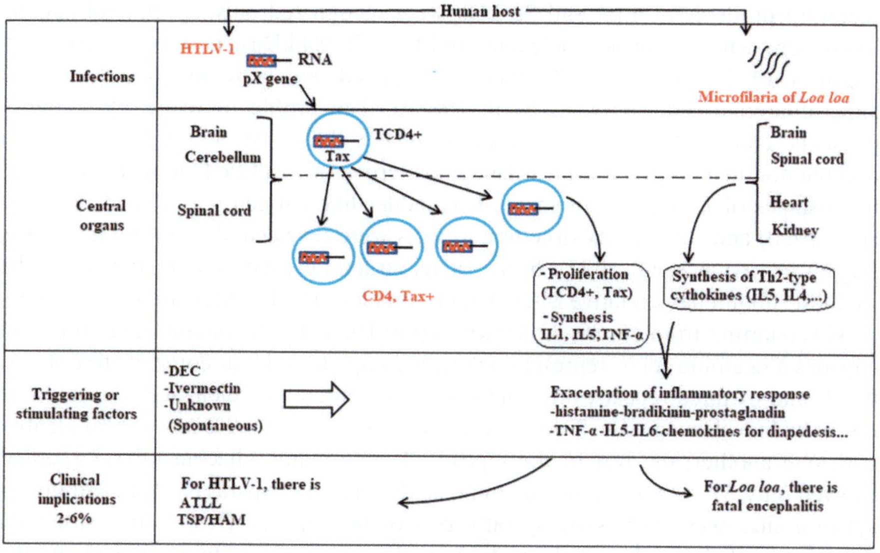

Fig. 2 Possible influences of coinfection of *L. loa* and HTLV-1 on the immune response (Dieki R 2023). In the human host cell, HTLV1 interacts with heparan sulfate proteoglycans (HSPG)/ neuropilin-1 (NRP-1) and GLUT-1 to penetrate human CD4 T cells. The gene *pX* coding for the Tax protein starts secreting this protein, which results in dysregulation of the cell cycle, survival, proliferation of T cells, cytokine, and cytokine receptor, driving the proliferation of multiple leukemic clonal CD4T^{+}Tax cells and synthesis of IL-1, IL-5, and TNF-α. In an environment with *L. loa* hypermicrofilaremia, there is an excessive production of IL-5, IL-4, IL-10, and IFN-γ. The combination of both environments leads to an inflammatory response (production of histamines, bradykinins, prostaglandins, TNF-α, IL-5, IL-6, and chemokines for diapedesis). This may be followed by serious clinical signs such as HTLV-1-associated myelopathy/tropical spastic paraparesis (HAM/TSP), adult T cell leukemia/lymphoma (ATLL), or fatal encephalitis with *L. loa*

ranging from 8 to 250 kDa (Akue et al. 1998); however, their nature is not well known. A preliminary analysis of a *L. loa* adult surface antigens revealed the existence of a major 29–31-kDa antigen, mostly reacting with IgG from amicrofilaremic individuals, but poorly with those from microfilaremic individuals. This molecule binds concanavalin A, was sensitive to proteases, and its antigenicity was resistant to heat but sensitive to periodate and endo-beta-*N*-acetylglucosaminidase H. These observations suggest that it is a glycoprotein containing mannose and *N*-acetylglucosamine residues and that the carbohydrate moiety is important for antibody binding (Egwang et al. 1988a). These observations show that the molecule is a glycoprotein that seems to be an N-linked glycoprotein homologous to the molecule from lymphatic filaria (Devaney 1987). There are other surface molecules with sizes of 14.5, 17, 21, 23, 34, 58, and 86 kDa, which are weakly recognized by the IgG of infected people. A similar analysis was made of microfilariae and revealed the existence of a 23-kDa immunodominant antigen. This antigen was neither glycosylated nor had an interchain disulfide linkage but reacted mostly against IgG from amicrofilaremic individuals. Additionally, excretory-secretory products of 14.4 and 33 kDa were recognized only by amicrofilaremic loiasis sera, whereas surface molecules of 40 and 75–84 kDa and excretory-secretory products of 28 and 67 kDa were recognized by both microfilaremic and amicrofilaremic sera (Egwang et al. 1988b). The profile of excretory-secretory products consisted of molecules of 14.4–198 kDa. At a somatic level, an immunodominant antigen with a ladder pattern was identified in all of the three main stages of *L. loa* development. This ladder-like antigen was cloned (Egwang et al. 1990) and some of its structural aspects were revealed. A 1764-bp insert was sequenced (Ajuh et al. 1995). Some of its structural aspects were revealed. The smallest subunit has a relative M(r) of approximately 15 kDa. Increased each time by 15 kDa (starting from 15, 30, 45, 60 kDa up to 105 kDa). It contains no introns and encodes 588 amino acids, representing one incomplete and four complete repeats. At the 3′ end of three repeats, there are consensus proteolytic cleavage sites and one repeat had no cleavage site. Two perfect repeats showed a 93.9% amino acid identity with one another; the rest of the repeats, despite being adjacent to one another, showed only 31–42% of identical amino acids. Putative asparagine N-linked glycosylation sites were expressed by only two of the repeats. Despite this structural diversity, each *L. loa* repeat showed a homology to *Ascaris suum* allergen and the homolog protein described in *Dirofilaria immitis* and *Brugia pahangi*. The repeat 3 design as r3 has been used for molecular diagnostics (Touré et al. 1997). This repeat is specific for *L. loa* and its antigenicity has been proven as a recombinant molecule (Azzibrouck et al. 2010). Cross-reactive antigens are widespread among filarial parasites. A gene was amplified in *L. loa* DNA using *B. pahangi* 21-kDa gene primers. The resulting amplicon was sequenced and translated to a protein, which generated a 17.225-kDa polypeptide. The polypeptide had a pI of 5.80, characterized by a valine-rich region, a leucine zipper, and phosphorylation, myristylation, and glycosylation sites. The hydropathicity plot showed alternate hydrophilic and hydrophobic regions. The alignment of this polypeptide with the *L. loa* hypothetical protein, with *B. pahangi*, *B. malayi*, and *W. bancrofti* showed between 46% and 47% homology (Akue et al. 2020). Cross-reactivity of antigens is widespread, even with other organisms. The major disadvantage of an application of these antigens is thus the lack of specificity of diagnostic tests used to identify *L. loa* parasites. However, some cross-reactive antigens may be useful for protective immunity. In this regard, the amplification of probable homologous ALT, and chitinase gene in *L. loa* genomic DNA, may be useful (Dieki 2023) for vaccine strategies.

5 Conclusion

The parasitological and clinical expression of *L. loa* displays a variety of immunological mechanisms underlying differences of reactivity against different parasite antigens. These mechanisms are complex, as is the regulation of immune response in loiasis. However, the occurrence of anergy characterized by diminished cell proliferation but increased IgG4 production is obvious. Furthermore, it is clear that protective immunity exists through concomitant immunity, manifested by elimination of the larval stage (microfilaria and L3). Although antigens have been identified, more research is still needed to determine their nature and structure. Therefore, further studies should be conducted for a better understanding of immune regulation and characterization of antigens that may be targeted for vaccines, diagnostics, or drug development.

Acknowledgment We thank Dr. Luzia Veletzky for corrections. The work on *loa loa* was supported by International center for Medical research of Franceville (CIRMF). CIRMF is sponsored by Total Gabon and Gabonese state.

References

Akue JP, Hommel M, Devaney E (1998) IgG subclass recognition of Loa loa antigens and their correlation with clinical status in individuals from Gabon. Parasite Immunol 20(8):387–393. https://doi.org/10.1046/j.1365-3024.1998.00172.x

Akue JP, Eyang-Assengone ER, Dieki R (2020) Immunological and molecular similarities between human filarial *Loa loa* and *Brugia pahangi* antigens. EC Microbiol 16(1):1–14

Akue JP, Morelli A, Moukagni R, Moukana H, Azzibrouck G (2003) Parasitological and immunological effects induced by immunization of Mandrillus sphinx against the Human filarial Loa loa using infective stage larvae irradiated at 40kRAD. Parasite 10(3):263–268. https://doi.org/10.1051/parasite/2003103263

Akue JP, Devaney E (2002) Transmission intensity affects both antigen-specific and nonspecific T-cell proliferative responses in Loa loa infection. Infect Immun 70(3):1475–1480. https://doi.org/10.1128/IAI.70.3.1475-1480.2002

Akue JP (1995) Correlation between the antibody response and resistance to Loa loa infection in Gabon. PhD these, University of Liverpool, p 292

Ajuh PM, Akue JP, Boutin P, Everaere S, Egwang TG (1995) Loa loa: structural diversity of a 15-kDa repetitive antigen. Exp Parasitol 81(2):145–153. https://doi.org/10.1006/expr.1995.1103

Azzibrouck GB, Akue JP, Lenoble DR (2010) Production and immunological characterization of a recombinant subunit of a Loa loa polyprotein antigen. Parasitology 137(7):1119–1128. https://doi.org/10.1017/S0031182009991740. Epub 2010 May 5

Ateba-Ngoa U, Mombo-Ngoma G, Zettlmeissl E, van der Vlugt LEPM, de Jong S et al (2014) CD4+ CD25hiFOXP3+ cells in cord blood of neonates born from filaria infected mother are negatively associated with CD4+ Tbet+ and CD4+ RORct + T cells. PLoS One 9(12):e114630. https://doi.org/10.1371/journal.pone.0114630

Baize S, Wahl G, Soboslay PT, Egwang TG, Georges AJ (1997) T helper responsiveness in human *Loa loa* infection; defective specific proliferation and cytokine production by $CD4^+$ T cells from microfilaraemic subjects compared with amicrofilaraemics. Clin Exp Immunol 108(2):272–278. https://doi.org/10.1046/j.1365-2249.1997.d01-1010.x

Chunda VC, Ritter M, Bate A, Gandjui NVT, Esum ME, Fombad FF, Njouendou AJ, Ndongmo PWC, Taylor MJ, Hoerauf A, Layland LE, Turner JD, Wanji S (2020) Comparison of immune responses to *Loa loa* stage-specific antigen extracts in Loa loa-exposed BALB/c mice upon clearance of infection. Parasit Vectors 13:51. https://doi.org/10.1186/s13071-020-3921

Devaney E (1987) Preliminary studies on the characterization of the Mr 30,000 surface antigen of Brugia pahangi. Parasite Immunol 9(3):401–405. https://doi.org/10.1111/j.13653024.1987.tb00518.x

Dieki R, Nsi-Emvo E, Akue JP (2022) The human filaria *Loa loa*: update on diagnostics and immune response. Res Rep Trop Med 2022:13

Dieki R (2023) Analyse de la réponse immunochimique et moléculaire des antigènes de la larve infectante de la filaire humaine (*Loa loa*). PhD thesis, USTM, p 123

Djikeussi ED, Akue JP (2014) Age-dependent prevalence of Loa loa amicrofilaremia and microfilaremia status as defined by two markers: microfilaria and specific IgG4. Afr J Biotechnol 13(4):593–597. https://doi.org/10.5897/AJB09.645. ISSN 1684-5315. ©2014 Academic Journals. http://www.academicjournals.org/AJB

Drame PM, Montavon C, Pion SD, Kubofcik J, Fay MP, Nutman TB (2016) Molecular epidemiology of blood-borne human parasites in a *Loa loa*-, *Mansonella perstans*-, and *Plasmodium falciparum*-endemic region of Cameroon. Am J Trop Med Hyg 94(6):1301–1308. https://doi.org/10.4269/ajtmh.15-0746

Egwang TG, Akué JP, Dupont A, Pinder M (1988a) The identification and partial characterization of an immunodominant 29-31 kilodalton surface antigen expressed by adult worms of the human filaria Loa loa. Mol Biochem Parasitol 31(3):263–272. https://doi.org/10.1016/0166-6851(88)90156-9

Egwang TG, Dupont A, Akué JP, Pinder M (1988b) Biochemical and immunochemical characterization of surface and excretory-secretory antigens of Loa loa microfilariae. Mol Biochem Parasitol 31(3):251–261. https://doi.org/10.1016/0166-6851(88)90155-7

Egwang TG, Pinder M, Akue JP (1990) Loa loa: identification of genomic DNA clones expressing recombinant antigens. Exp Parasitol 70(4):490–493. https://doi.org/10.1016/0014-4894(90)90135-y

Hussain R, Poindexter RW, Ottesen EA (1992) Control of allergic reactivity in human filariasis. Predominant localization of blocking antibody to the IgG4 subclass. J Immunol 148(9):2731–2737

Hartmann S, Lucius R (2003) Modulation of host immune responses by nematode cystatins. Int J Parasitol 33(11):1291–1302

Haapasalo K, Meri T, Jokiranta TS (2009) *Loa loa* Microfilariae evade complement attack in vivo by acquiring regulatory proteins from host plasma. Infect Immun 77(9):3886–3893. https://doi.org/10.1128/IAI.01583-08. Epub 2009 Jun 15. PMID: 19528206; PMCID: PMC2738024

Leroy E, Baize S, Wahl G, Egwang TG, Georges AJ (1997) Experimental infection of a nonhuman primate with Loa loa induces transient strong immune activation followed by peripheral unresponsiveness of helper T cells. Infect Immun 65(5):1876–1882

Lemnui B, Mahamat O, Ngum NH, Dorette NA, Akenji TK (2020) Transplacental transmission of filarial infections and its impact on oxidative stress and immune responses during foetal life in Bamenda, North West Cameroon. Int J Trop Dis Health 41(13):26–38. https://doi.org/10.9734/ijtdh/2020/v41i1330345

Mogoung-Wafo AE, Nana-Djeunga HC, Domche A, Fossuo-Thotchum F, Bopda J, Mbickmen-Tchana S, Djomo-Kamga H, Kamgno J (2019) Prevalence and intensity of Loa loa infection over twenty-three years in three communities of the Mbalmayo health district (Central Cameroon). BMC Infect Dis 19:146. https://doi.org/10.1186/s12879-019

Njambe Priso GD, Lissom A, Ngu LN, Nji NN, Tchadji JC, Tchouangueu TF, Ambada GE, Ngane CSS, Dafeu BL, Djukouo L, Nyebe I, Magagoum S, Ngoh AA, Herve OF, Garcia R, Gutiérrez A, Okoli AS, Esimone CO, Njiokou F, Park CG, Waffo AB, Nchinda GW (2018) Filaria specific antibody response profiling in plasma from anti-retroviral naïve Loa loa microfilaraemic HIV-1 infected people. BMC Infect Dis 18(1):160. https://doi.org/10.1186/s12879-018-3072-2

Nutman TB, Miller KD, Mulligan M, Ottesen EA (1986) *Loa loa* infection in temporary residents of endemic regions: recognition of a hyperresponsive syndrome with characteristic clinical manifestations. J Infect Dis 154(1):10–18. https://doi.org/10.1093/infdis/154.1.10

Ricciardi A, Nutman TB (2021) IL-10 and its related superfamily members IL-19 and IL-24 provide parallel/redundant immune-modulation in Loa loa infection. J Infect Dis 223(2):297–305. https://doi.org/10.1093/infdis/jiaa347

Touré FS, Bain O, Nerrienet E, Millet P, Wahl G, Toure Y, Doumbo O, Nicolas L, Georges AJ, McReynolds LA, Egwang TG (1997) Detection of Loa loa-specific DNA in blood from occult-infected individuals. Exp Parasitol 86(3):163–170. https://doi.org/10.1006/expr.1997.4168

Van Hoegaerden M, Chabaud B, Akue JP, Ivanoff B (1987) Filariasis due to *Loa loa* and *Manspnella perstans*: distribution in the region of Okondja, Haut-Ogooué Province, Gabon, with parasitological and serological follow-up over one year. Trans R Soc Trop Med Hyg 81 (3):441–446. https://doi.org/10.1016/0035-9203(87)90163-5

The *Loa loa* Genome and Potential Immunological and Therapeutic Molecular Targets with Medicinal Plants

Roland Dieki, Line Edwige Mengome, and Jean Paul Akue

Abstract

The sequence of the whole *Loa loa* genome has provided new insights and a better understanding of the parasite and its life span in the human body. This parasite, which does not have a symbiont, can survive by using its one metabolism pathway. The *L. loa* genome contains genes for many crucial enzymes, including members of the GPCR family and protein kinases. Genes involved in immune evasion are also present in this genome. In this report, we dissect some of the metabolic pathways that could lead to new diagnostics, vaccines, and drug development (including medicinal plants).

Keywords

Loa loa · Genome · Chemotherapy pathways · Vaccine · Medicinal plants

1 Introduction

Although the the use of genomics to validate plant-derived drugs has gained interest, to date it has not led to much; however, the genome has revolutionized diagnostics and has been useful in demonstrating that *Loa loa* does not contain *Wolbachia* and cannot be treated with doxycycline.

Early studies identified genomic clones making proteins (Egwang et al. 1990). Furthermore, a species-specific repetitive DNA was cloned (Klion et al. 1991) and

R. Dieki · J. P. Akue (✉)
Neglected Parasitosis, CIRMF, Franceville, Gabon

L. E. Mengome
Institut de Pharmacopée et de Médecine Traditionnelle (IPHAMETRA – CENAREST), Libreville, Gabon

J. P. Akue (ed.), *Loa loa: Latest Advances in Loiasis Research*,
https://doi.org/10.1007/978-3-031-49450-5_8

has become widely used for diagnostics. This was improved by a qPCR developed with sequence tags specific for *Loa loa* microfilariae, which can detect up to 0.1 pg of genomic DNA (Fink et al. 2011) and is useful for quantitation of microfilariae necessary to avoid side effects after drug administration. Based on the fact that the biological mechanisms that underpin the presence or absence of Wolbachia are not well known, comparison of the genomes of Wolbachia-dependent and Wolbachia-independent species did not show any discernable differences between filarial mitochondrial genome from wolbachia-dependent or -independent (McNulty et al. 2012). The genome of *L. loa* was first sequenced in 2013 (Desjardin et al. 2013). This ushered in many new avenues of research, especially related to the identification of biomarkers, possible pathways for new drug development, and potential antigens for use in vaccine development. The potentially fatal complications of drugs currently used for the treatment of *L. loa* led to screening of many molecules in plants used for traditional medicine for their activity in the inhibition of *L. loa* gene activity and/or pathways used by the parasite to survive.

2 General Overview of *L. loa* Genome Structure

Loa loa is one of the less studied among human pathogenic filariae. However, recent interest in this filaria has increased in the setting of fatal adverse events during mass drug administration programs using ivermectin for onchocerciasis control. The *L. loa* genome was published in 2013 (Desjardins et al. 2013), and the comparison of this genome with other nematode genomes has provided information about molecules and pathways specific to *L. loa*. The *L. loa* genome size is 91.4 Mb and composed of six chromosomes, with five being autosomes and one sex chromosome. Based on 70% RNA sequencing, the genome codes for 14,907 genes with specific functions. It is interesting to note that *L. loa* can synthesize metabolites that other filarial cannot without the assistance of the symbiont *Wolbachia*; this symbiont does not exist in *L. loa* (Buttner et al. 2003; Grobusch et al. 2003; McGarry et al. 2003).

2.1 Genes Implicated in Important Metabolic Pathways

Numerous genes involved in metabolic pathways have been identified in the *L. loa* genome. Many of the metabolic pathways that *Wolbachia* was thought to provide do not exist in *L. loa* (Desjardins et al. 2013).

For example, the *L. loa* genome encodes multiple enzymes involved in B6 metabolism, including pyridoxamine 5′-phosphate oxidase, which is essential for the synthesis of vitamin B6, and five copies of the gene coding the second enzyme (pyridoxal 5′-phosphate synthase) implicated in vitamin B6 uptake and recycling. Other enzymes likely facilitate migration of the worms through host tissues, such as those coding for hyaluronidases, proteins that facilitate tissue penetration, and chemoreceptor genes that likely enhance host–parasite interactions. The presence of trehalase domains involved in trehalose degradation have also been reported.

Trehalose protects nematodes from environmental stress and has been implicated in the long life span of *C. elegans* (Pellerone et al. 2003; Honda et al. 2010). Several members of the transmembrane protein G family (GPCR) including Srab, Srx, Srsx, and Srw are conserved among nematodes, including *L. loa*. Finally, *L. loa* has the capacity to biosynthesize certain metabolites, such as FAD and glutathione, and to partially synthesize pyrimidines, unlike other filarial parasites that require the presence of the endosymbiont *Wolbachia* for these molecules (Foster et al. 2005).

2.2 Genes Implicated in Immune Escape

Genome sequencing revealed that *L. loa* has defense system genes that allow the parasite to escape the immune system both of humans and of the arthropod vector. Within this system, molecules of *L. loa* innate immunity, such as lectin type C, galectins, jacalins, and scavenger receptors, have been identified. The gene encoding lipopolysaccharide-binding protein is also present, as well as several mimics and antagonists of human cytokines and chemokines, including mimics of macrophage migration inhibitor (MIF), transforming growth factor-β and its receptor, a member of the interleukin (IL-16) family, an antagonist of IL-5 receptor, an interferon regulatory factor, a homolog of the suppressor of cytokine signaling (SOCS)-7, and two members of the chemokine-like family. Also present in the *L. loa* genome are genes coding for molecules such as serpins and cystatins, which interfere with antigen processing and presentation to T cells (Hartmann and Lucius 2003), the gene for indoleamine 2,3 dioxygenase (IDO), an immunomodulatory protein implicated in immune escape, and numerous genes from the *Wnt* family implicated in activation of immune modulation. Finally, genes whose sequences resemble human autoantigens have been identified and could indicate increased autoreactivity of *L. loa* antibodies compared to other parasites (Desjardins et al. 2013). However, it should be pointed out that there are no functional data for any of these genes found in the *L. loa* genome.

2.3 Genes for Potential Immunogenic Molecules

The *L. loa* genome has several immunogenic domains responsible for cellular adhesion. Recently, genes coding for antigens with vaccine potential were identified in *L. loa* (Fig. 1). These include orthologs of the genes for the stage-specific *Brugia malayi* candidate vaccine antigens, *Abundant Larval Transcript-1* and *ATL-2* (*ATL-1* and *ATL-2*). An ortholog of the *Brugia chitinase* has also been found in *L. loa* (Dieki 2023). These antigens have proven their vaccine potential in experimental models of *Brugia*. Immunization of susceptible *Meriones unguiculatus* with *ALT-1* induced a 76% reduction in parasite survival (Gregory et al. 2000). Similarly (Ramachandran et al. 2004), immunization of mice with recombinant *ALT-2* or DNA-encoding *ALT-*

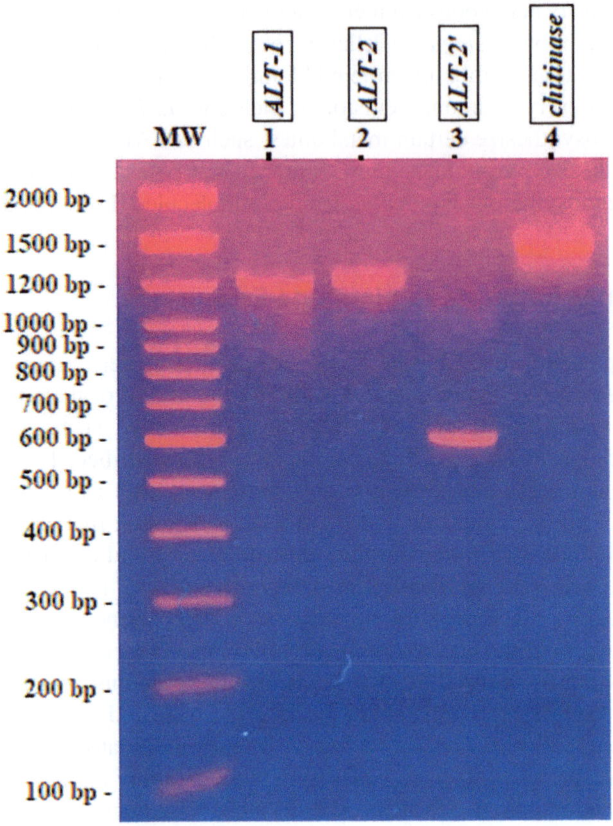

Fig. 1 Amplification of three genes of *L. loa* L3. The homologs of *ALT-1*, *ALT-2,* and *chitinase* were amplified in *L. loa* using *B. malayi* primers (Dieki 2023). The results were analyzed after electrophoresis on an agarose gel (1.5%) and visualized under UV light. The amplified gene is indicated at the top of each band along with its corresponding size, in comparison with the standard molecular weight markers (PM) in the left end of the figure: **1:** *ALT-1* (≈1200 bp), **2:** *ALT-2* (±1200 bp), and **4:** *chitinase* (≈1500 bp). However, amplification of *ALT-2* generated an additional band corresponding roughly to half of the main band, i.e., **3:** *ALT-2′* (±600 bp)

2 produced a significant level of specific antibodies in mice. Moreover, specific antibody levels were found to be increased in individuals endemic control compared with those with filarial infection (Dissanayake et al. 1995). In parallel experiments with *chitinase* (Adam et al. 1996) and SXP1 antigen, partial protection against *Brugia* microfilariae but not adult worms was observed for *chitinase,* whereas SXP1 protected against microfilariae and adult worms. Filarial myosin showed no protection (Wang et al. 1997). It should be noted that none of the vaccine models is a loiasis model.

3 Potential Biotherapeutic Pathways

One of the biggest problems in loiasis is the lack of a safe and effective macrofilaricidal therapy. Both ivermectin and DEC have marked microfilaricidal activity that impacts safety, and albendazole is impractical at the doses required for efficient killing of adult worms. The *L. loa* genome affords at least two potential approaches to this problem: (1) the G protein-coupled receptor (GPCR) family and (2) protein kinases. The GPCR family is composed of classes A (rhodopsin), B [secretin (B1) and adhesion (B2)], C (glutamate), and F (Frizzled). The availability of the 3D structure and the bioinformatic simulations have helped to reveal the conformations of their inactive and active forms (Yang et al. 2021). For this reason, it is possible to predict how GPCR binds to numerous ligands and how they transmit different signals to generate specific physiological functions. This information may help in designing antagonists to block GPCR activity. Molecules of interest can then be screened via a DNA-encoded library (DEL) (Song and Li 2021) or affinity selection mass spectrometry (MS) (van Breemen 2020). Proteins kinases provide excellent targets for drugs to fight against different infections. This assertion is substantiated by the fact that an inhibitor of the tyrosine kinase imatinib is capable of killing schistosomes (Beckmann and Grevelding 2010) and all stages of *Brugia* in vitro (Beckmann and Grevelding 2010); similar activity was shown with *L. loa* adult worms and microfilariae (O'Connell and Thomas B. Nutman 2017; O'Connell et al. 2017). The kinase TTK (Hogg et al. 1994), which is necessary for meiosis in eucaryotic cells, as well as six other potential protein kinase targets are also present in *L. loa*. In general, protein kinases selectively modify other proteins by covalently adding phosphates to them (phosphorylation). There are two major families of protein kinases: (a) the serine/threonine kinases and (b) the tyrosine kinases. One interesting protein kinase is the AMP-activated protein kinase (AMPK), which is important for muscle function. Mice lacking AMPK experience rapid fatigue (O'Neill et al. 2011; Lantier et al. 2014), while chronic activation of AMPK improves muscle function (Röckl et al. 2007; Narkar et al. 2008). The activity of this protein kinase could be exploited in *L. loa* by using a molecule that affects AMPK activity, provided that the degree of homology between the *L. loa* AMPK and the human counterpart enables the design of a specific inhibitor of the parasite AMPK. It is known that there are drugs that activate and inhibit AMPK in humans as in the case of cancer or diabetes type 2 (Hardie 2013). Screening of activity could be performed using previously developed techniques (Davies et al. 1989; Chen et al. 2000; Guigas et al. 2006; Annis et al. 2007) (Fig. 2).

4 Interaction Between Therapeutic Molecules from Medicinal Plants and the *Loa loa* Parasite

Medicinal plants have been used for a long time. In fact, approximately 60% of the world's population relies on plants for primary healthcare needs (WHO 2011). The WHO Traditional Medicine Centers indicated that 80% of 122 compounds identified

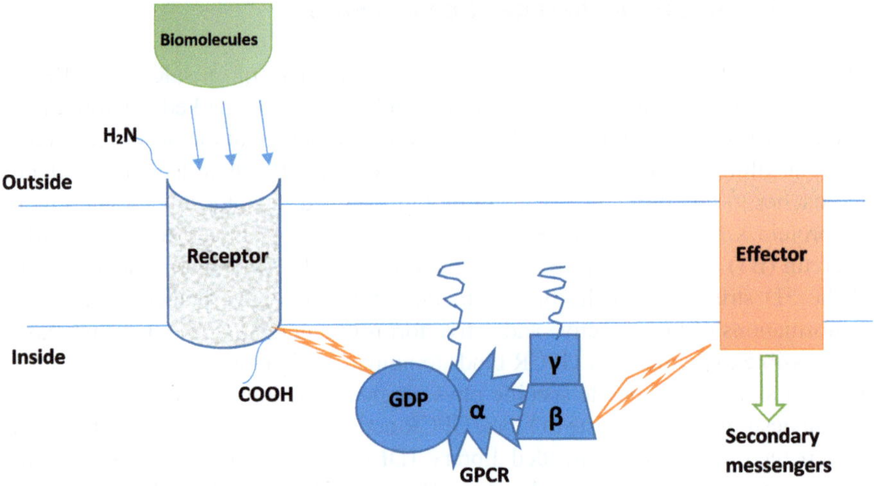

Fig. 2 Illustration of targeting GPCR pathways: the receptor transmits information to intracellular G-proteins, which in turn activate or inhibit intracellular effectors. As long as the receptor is activated by its ligand, and as long as the system does not undergo desensitization, the cycle of exchange of GDP for GTP continues

were used for the same or related ethnomedical purposes and were derived from only 94 plants species (Cragg and Newman 2013). Many plants have shown activity against *L. loa*, and phytochemical studies have shown that secondary metabolites, such as alkaloids, polyphenols (flavonoids, tannins, coumarins), sugars, sterols, terpenoids, saponins, also have an effect (Mengome et al. 2010, 2018). Indeed, a Daflon-micronized purified flavonoid fraction of *Rutaceae aurantiae* has been shown to have activity in the treatment of filarial edema (Das et al. 2003). Saponin has a detergent action on the biomembrane, and synthetic alkaloids have been reported to have antifilarial activity in vivo (Singh et al. 1997). Sugar-based molecules have shown macrofilaricidal activity (Chatterjee et al. 1992), and two compounds (voacangine and vocamine) isolated from the stem bark of *Vocanga africana* inhibited the motility of *L. loa* microfilariae and adult worms of *Onchocerca ochengi* with half maximal inhibition concentrations (IC_{50}) of 5.49 μM and 2.49 μM for *L. loa* and 9.07 μM and 3.45 μM for *Onchocerca ochengi* with voacangine and vocamine, respectively (Borakaeyabe et al. 2015). The IC_{50} of methanol extracts for *Petersianthus macrocarpus* bark, for *Piptadeniastrum africanum*, for *Vernonia conferta* bark, and for *Lophira alata* bark on *L. loa* microfilariae were 0.22, 1.082, 3.78, and 5.29, respectively, compared to standard drugs, such as DEC (0.38 μg/m) and IVM (32.74 μg/ml). These molecules as well as others yet to be identified may, in combination with information derived from bioinformatics analysis of the *L. loa* genome, result in the design of a novel macrofilaricide (van Breemen 2020; Muchiri and van Breemen 2021; Kodadek et al. 2019).

5 Conclusion

The genome reveals many important genes and metabolism pathways that could be explored to find sensitive, specific, and cheap point-of-care diagnostic tools. Genes homologous to those found to induce protective immunity in lymphatic filariasis seem to be present in *L. loa* and could lead to pan-filarial vaccine development. Finally, although finding a macrofilaricide remains a challenging task in filariasis, combining knowledge of *L. loa* molecular structure and information from studies of medicinal plants may result in the design of a new drug capable of killing adult worms.

Acknowledgment We thank Amy Klion for English and useful comments. The work on *loa loa* was supported by International center for Medical research of Franceville (CIRMF). CIRMF is sponsored by Total Gabon and Gabonese state.

References

Adam R, Kaltmann B, Rudin W, Friedrich T, Marti T, Lucius R (1996) Identification of chitinase as the immunodominant filarial antigen recognized by sera of vaccinated rodents. J Biol Chem 271: 1441–1447

Annis A, Chuang C-C, Nazef N (2007) ALIS: an affinity selection–mass spectrometry system for the discovery and characterization of protein–ligand interactions. In: Mannhold R, Kubinyi H, Folkers G, Wanner KT, Höfner G (eds) Mass spectrometry in medicinal chemistry, pp 297–305. ISSN 0166-6851. https://doi.org/10.1016/0166-6851(91)90098-Q

Beckmann S, Grevelding CG (2010) Imatinib has a fatal impact on morphology, pairing stability and survival of adult Schistosoma mansoni in vitro. Int J Parasitol 40(5):521–526

Borakaeyabe SB, Mbah JA, Cho-Ngwa F, Metuge JA, Efange MN, S. (2015) Isolation and characterization of flaricidal compounds from the stem bark of Vocanga africana, a plant used in the traditional treatment of onchocerciasis in Cameroon. J Med Plant Res 9:471–478

Büttner DW, Wanji S, Bazzocchi C, Bain O, Fischer P (2003) Obligatory symbiotic Wolbachia endobacteria are absent from Loa loa. Filaria J 2(1):10. https://doi.org/10.1186/1475-2883-2-10

Chatterjee RK, Fatma N, Murthy PK, Sinha P, Kulshrestha DK, Dhawan BN (1992) Macrofilaricidal activity of the stembark of Streblus asper and its major active constituents. Drug Dev Res 26:67–78

Chen ZP, McConell GK, Michell BJ, Snow RJ, Canny BJ, Kemp BE (2000) AMPK signaling in contracting human skeletal muscle: acetyl-CoA carboxylase and NO synthase phosphorylation. Am J Physiol Endocrinol Metab 279(5):E1202–E1206

Cragg GM, Newman DJ (2013) Natural products: a continuing source of novel drug leads. Bioch Biophysica Acta Gen Subj 1830:3670–3695

Das LK, Subramanyam Reddy G, Pani SP (2003) Some observation on the effect of Daflon (micronized purified flavonoid fraction of Rutacea aurantiae) in Bancroftian filarial lymphoedema. Filarial J 2:1–9

Davies SP, Carling D, Hardie DG (1989) Tissue distribution of the AMP-activated protein kinase, and lack of activation by cyclic-AMP-dependent protein kinase, studied using a specific and sensitive peptide assay. Eur J Biochem 186(1–2):123–128

Desjardins CA, Cerqueira GC, Goldberg JM, Dunning Hotopp JC, Haas BJ, Zucker J, Ribeiro JM, Saif S, Levin JZ, Fan L, Zeng Q, Russ C, Wortman JR, Fink DL, Birren BW, Nutman TB (2013) Genomics of Loa loa, a Wolbachia-free filarial parasite of humans. Nat Genet 45(5):495–500

Dieki R (2023) Analyse de la réponse immunochimique et moléculaire des antigens de la larve infectante de la filaire humaine (*Loa loa*). [Thèse de doctorat]. Ecole Doctorale des Sciences Fondamentales et Appliquées (ED-FSA). Université des Sciences et Techniques de Masuku (USTM), p 122

Dissanayake S, Perler FB, Xu M et al (1995) Differential recognition of microfilarial chitinase, a transmission-blocking vaccine candidate antigen, by sera from patients with Brugian and Bancroftian filariasis. Am J Trop Med Hyg 53:289–294

Egwang T, Pinder M, Akue JP (1990) Loa loa: identification of genomic DNA clones expressing recombinant antigens. Exp Parasitol 70(4):490–493. https://doi.org/10.1016/0014-4894(90)90135-y

Fink DL, Kamgno J, Nutman TB (2011) Rapid molecular assays for specific detection and quantitation of Loa loa microfilaremia. PLoS Negl Trop Dis 5(8):e1299. https://doi.org/10.1371/journal.pntd.0001299. Epub 2011 Aug 30

Foster J, Ganatra M, Kamal I, Ware J, Makarova K, Ivanova N, Bhattacharyya A, Kapatral V, Kumar S, Posfai J, Vincze T, Ingram J, Moran L, Lapidus A, Omelchenko M, Kyrpides N, Ghedin E, Wang S, Goltsman E, Joukov V, Ostrovskaya O, Tsukerman K, Mazur M, Comb D, Koonin E, Slatko B (2005) The *Wolbachia* genome of *Brugia malayi*: endosymbiont evolution within a human pathogenic nematode. PLoS Biol 3(4):e121

Gregory WF, Atmadja AK, Allen JE, Maizels RM (2000) The *abundant larval transcript-1* and -*2* genes of *Brugia malayi* encode stage-specific candidate vaccine antigens for filariasis. Infect Immun 68(7):4174–4179

Grobusch MP, Kombila M, Autenrieth I, Mehlhorn H, Kremsner PG (2003) No evidence of Wolbachia endosymbiosis with Loa loa and Mansonella perstans. Parasitol Res 90(5):405–408. https://doi.org/10.1007/s00436-003-0872-z. Epub 2003 May 14

Guigas B, Bertrand L, Taleux N, Foretz M, Wiernsperger N, Vertommen D, Andreelli F, Viollet B, Hue L (2006) 5-Aminoimidazole-4-carboxamide-1-beta-D-ribofuranoside and metformin inhibit hepatic glucose phosphorylation by an AMP-activated protein kinase-independent effect on glucokinase translocation. Diabetes 55(4):865–874

Hardie DG (2013) AMPK: a target for drugs and natural products with effects on both diabetes and cancer. Diabetes 62(7):2164–2172. https://doi.org/10.2337/db13-0368

Hartmann S, Lucius R (2003) Modulation of host immune responses by nematode cystatins. Int J Parasitol 33(11):1291–1302

Hogg D, Guidos C, Bailey D, Amendola A, Groves T, Davidson J, Schmandt R, Mills G (1994) Cell cycle-dependent regulation of the protein kinase TTK. Oncogene 1:89–96

Honda Y, Tanaka M, Honda S (2010) Trehalose extends longevity in the nematode Caenorhabditis elegans. Aging Cell 9(4):558–569

Klion AD, Raghavan N, Brindley PJ, Nutman TB (1991) Cloning and characterization of a species-specific repetitive DNA sequence from Loa loa. Mol Biochem Parasitol 45(2)

Kodadek T, Paciaroni NG, Balzarini M, Dickson P (2019) Beyond protein binding: recent advances in screening DNA-encoded libraries. Chem Commun (Camb) 55(89):13330–13341

Lantier L, Fentz J, Mounier R, Leclerc J, Treebak JT, Pehmøller C, Sanz N, Sakakibara I, Saint-Amand E, Rimbaud S, Maire P, Marette A, Ventura-Clapier R, Ferry A, Wojtaszewski JF, Foretz M, Viollet B (2014) AMPK controls exercise endurance, mitochondrial oxidative capacity, and skeletal muscle integrity 28(7):3211–3224

McGarry HF, Pfarr K, Egerton G, Hoerauf A, Akue JP, Enyong P, Wanji S, Kläger SL, Bianco AE, Beeching NJ, Taylor MJ (2003) Evidence against Wolbachia symbiosis in Loa loa. Filaria J 2(1):9. https://doi.org/10.1186/1475-2883-2-9

McNulty SN, Mullin AS, Vaughan JA et al (2012) Comparing the mitochondrial genomes of *Wolbachia*-dependent and independent filarial nematode species. BMC Genomics 13:145. https://doi.org/10.1186/1471-2164-13-145

Mengome LE, Akué JP, Souza A, Feuya Tchoua GR, Nsi EE (2010) In vitro activities of plant extracts on human *Loa loa* isolates and cytotoxicity for eukaryotic cells. Parasitol Res 107:643–650

Mengome LE, Mewono L, Aboughe-Angone S (2018) Are medicinal plants the future of *Loa loa* treatment? Pharmacogn Rev 12(23):133–137

Muchiri RN, van Breemen RB (2021) Affinity selection-mass spectrometry for the discovery of pharmacologically active compounds from combinatorial libraries and natural products. J Mass Spectrom 56(5):e4647

Narkar VA, Downes M, Yu RT, Embler E, Wang YX, Banayo E, Mihaylova MM, Nelson MC, Zou Y, Juguilon H, Kang H, Shaw RJ, Evans RM (2008) AMPK and PPARdelta agonists are exercise mimetics. Cell 134(3):405–415

O'Connell EM, Nutman TB (2017) Reduction of Loa loa microfilaremia with imatinib – a case report. N Engl J Med 377(21):2095–2096. https://doi.org/10.1056/NEJMc1712990

O'Connell EM, Kamenyeva O, Lustigman S, Bell A, Nutman TB (2017) Defining the target and the effect of imatinib on the filarial c-Abl homologue. PLoS Negl Trop Dis 11(7):e0005690. https://doi.org/10.1371/journal.pntd.0005690

O'Neill HM, Maarbjerg SJ, Crane JD, Jeppesen J, Jørgensen SB, Schertzer JD, Shyroka O, Kiens B, van Denderen BJ, Tarnopolsky MA, Kemp BE, Richter EA, Steinberg GR (2011) AMP-activated protein kinase (AMPK) beta1beta2 muscle null mice reveal an essential role for AMPK in maintaining mitochondrial content and glucose uptake during exercise. Proc Natl Acad Sci U S A 108(38):16092–16097

Pellerone FI, Archer SK, Behm CA, Grant WN, Lacey MJ, Somerville AC (2003) Trehalose metabolism genes in Caenorhabditis elegans and filarial nematodes. Int J Parasitol 33(11): 1195–1206

Ramachandran S, Kumar MP, Rami RM, Chinnaiah HB, Nutman T, Kaliraj P, McCarthy J (2004) The larval specific lymphatic filarial ALT-2: induction of protection using protein or DNA vaccination. Microbiol Immunol 48(12):945–955

Röckl KS, Hirshman MF, Brandauer J, Fujii N, Witters LA, Goodyear LJ (2007) Skeletal muscle adaptation to exercise training: AMP-activated protein kinase mediates muscle fiber type shift. Diabetes 56(8):2062–2069

Singh RH, Singhal KC, Khan NU (1997) Antifilarial activity of *Mallotus philippensis* Lam. On *Setaria cervie* (Nematoda: Filaroidea) *in vitro*. Indian J Physiol Pharmacol 41:397–403

Song Y, Li X (2021) Evolution of the selection methods of DNA-encoded chemical libraries. Acc Chem Res 54(17):3491–3503

van Breemen RB (2020) Affinity selection-mass spectrometry: defining the bioactive compounds in complex mixtures of natural products and combinatorial libraries 18(1):18–23

Wang SH, Zheng HJ, Dissanayake S, Cheng WF, Tao ZH, Lin SZ, Piessens WF (1997) Evaluation of recombinant chitinase and SXP1 antigens as antimicrofilarial vaccines. Am J Trop Med Hyg 56(4):474–481

World Health Organization (2011) Global programme to eliminate lymphatic filariasis. A manual for National Elimination Programmes. Monitoring and Epidemiological Assessment of Mass Drug Administration, Geneva, pp 1–100

Yang D, Zhou Q, Labroska V, Qin S, Darbalaei S, Wu Y, Yuliantie E, Xie L, Tao H, Cheng J, Liu Q, Zhao S, Shui W, Jiang Y, Wang MW (2021) G protein-coupled receptors: structure- and function-based drug discovery. Signal Transduct Target Ther 6(1):7

MIX
Papier aus verantwortungsvollen Quellen
Paper from responsible sources
FSC® C105338

If you have any concerns about our products,
you can contact us on
ProductSafety@springernature.com

In case Publisher is established outside the EU,
the EU authorized representative is:
Springer Nature Customer Service Center GmbH
Europaplatz 3, 69115 Heidelberg, Germany

Printed by Libri Plureos GmbH
in Hamburg, Germany